D0617126

SOBER CURIOUS

SOBER CURIOUS

The Blissful Sleep, Greater Focus, Limitless Presence, and Deep Connection Awaiting Us All on the Other Side of Alcohol

RUBY WARRINGTON

HarperOne
An Imprint of HarperCollins*Publishers*

HarperOne

This book contains advice and information relating to health and interpersonal well-being. It is not intended to replace medical or psychotherapeutic advice and should be used to supplement rather than replace any needed care by your doctor or mental health professional. While all efforts have been made to ensure accuracy of the information contained in this book as of the date of publication, the publisher and the author are not responsible for any adverse effects or consequences that may occur as a result of applying the methods suggested in this book.

HarperCollins books may be purchased for educational, business, or sales promotional use. For information, please email the Special Markets Department at SPsales@harpercollins.com.

FIRST EDITION

Designed by Janet Evans-Scanlon

Library of Congress Cataloging-in-Publication Data
Names: Warrington, Ruby, author.
Title: Sober curious : the blissful sleep, greater focus, limitless presence, and deep connection awaiting us all on the other side of alcohol / Ruby Warrington.
Description: First edition. | New York, NY : HarperOne, [2018] | Includes bibliographical references.
Identifiers: LCCN 2018034065 (print) | LCCN 2018036100 (ebook) | ISBN 9780062869050 (e-book) | ISBN 9780062869036 (hardcover) | ISBN 9780062869043 (pbk.) | ISBN 9780062914569 (Intl.)
Subjects: LCSH: Temperance. | Drinking of alcoholic beverages—Psycho logical aspects. | Drinking of alcoholic beverages—Health aspects.
Classification: LCC HV5060 (ebook) | LCC HV5060 .W335 2018 (print) | DDC 616.86/106—dc23
LC record available at https://lccn.loc.gov/2018034065

19 20 21 22 23 LSC 10 9 8 7 6 5 4 3 2 1

To Cocktail Girl

CONTENTS

CONTENTS

INTRODUCTION

When I first got Sober Curious, one persistent question kept blinking into view, like a lighthouse on a stormy night:

Would life be better without alcohol?

This inquiry began as a conversation with my body before the words fully crystallized in my mind. On Sundays when my head hurt from the drinking. And not just my head, but the contents of my head. When my gut roiled, my tongue was furry with forgotten words, and even my *hair* felt hungover, greasy and crispy dry at the same time. Smelling of cigarettes and sour breath. Sometimes, on days like these, it felt like there was a hollow where my heart was supposed to be.

Another similar question had also been stalking me, on Tuesdays and Wednesdays and Thursdays, spent mainly, when I wasn't buckling under the stress of what had been my *dream job*, counting the hours until the weekend:

Is this really all there is?

Back then, I thought of myself as a moderate to heavy social drinker. Meaning I drank no more than most of the people I did my drinking with, and never (except on vacation) more than two nights in a row. But still far surpassing standard government guidelines as to what was "healthy." Seven units a week? I'm pretty sure I burned through those with a couple of cheeky mid-week pints of Stella.

It's impossible to pinpoint the precise moment these questions first began to demand answers. My memory is also fuzzy (funny that) about the circumstances leading to it. Was it something dramatic, like the time I fell, drunk, my friend and I each squished into a leg of an adult onesie, and bashed a bloody hole in my head? When instead of asking to be taken to the ER, I insisted that the best medicine would be a chunky slug of single malt whiskey? Or was it more mundane? One murky Monday morning too many hauling my woes to work like a sack of mildewed green potatoes. Since most of my Mondays used to feel that way, it's very hard to say.

But whenever it was, these first questions soon led to other questions. As if they'd all been hanging around, like reporters outside a court of law, waiting for the first to get my attention so they could have their turn.

Would I be happier without booze? More productive? Would I feel more confident? What would it be like to *never* have to face another deadline half hungover? Would I be thinner if I didn't drink? Look younger? Would I have less sex? More sex? Would the sex be better? Would I have anything to talk about at parties? Where would the glamour go? Would people think I was boring? Exactly *how* boring would I/life become?

Since you've picked up this book, I suspect these lines of inquiry may be familiar to you, too. No? Then how about these:

Why is alcohol so . . . *everywhere*? How come I feel like an outsider, a weirdo, sometimes a *problem*, if I say I don't drink? Why do I sometimes lie about why I'm not drinking? Where do I go to socialize without booze? How do I kick it with people who do drink if I'm not?

If you drink like I used to, maybe you've even been confronted with the big one: *Does the fact I'm asking all these questions make me . . . an alcoholic?*

Maybe.

I need to state up front that I am not a doctor, a brain scientist, or an addiction expert, and so it's really not for me to diagnose your personal drinking habits. But whichever sorry Sunday or miserable Monday the questioning first began for me, it soon lit the touch-paper on a radical reevaluation of my relationship to booze—an experiment that has expanded up and out to touch every part of my life, and which has meant I've spent most of the past decade seeking answers to these questions, and more.

Doing so has completely changed the way I drink and the way I think about drinking. Has shifted my entire perspective on the ways in which *we* drink, and the role alcohol plays in our relationships, our creativity, our happiness, and our society. As a result, I've also created a life for myself that is so exhilarating and rewarding, sometimes it feels like coming close to what they call *having it all.*

I have termed this questioning as getting Sober Curious. Maybe, for you, the questions are more focused on the possibilities of a life less sozzled. Ways to live your most vital life. What it might take to reverse the rules that somehow made drinking the socially acceptable thing to do. But in my book, if you have found yourself asking any or all of the above, then you are Sober Curious, too.

• • •

Since I'm not a doctor, a brain scientist, or an addiction expert, what exactly qualifies me to be writing a book like this? With twenty years working in journalism under my belt, I could tell you that it's my research skills and my nose for a lead. My well-honed ability to marry *A* with *B* to reach conclusion *C*.

But it begins with my telling you a little more about me.

I learned to drink at a late age, since I was a teetotaler all through college. But I got my first taste of alcohol at around age eight. Anyone? It's really not so shocking, if your parents, like mine, were of the mind that a child's curiosity sometimes deserves a grown-up response.

I have a vivid memory of "swimming" across my auntie's carpet, having sampled my first few sips of sherry. Then there was the picnic when, perhaps around age nine, I discovered the curious effects of grapefruit soda mixed with wine. I can still feel the giggles contained in the bubbles of this bouncy new drink tickling my throat.

My first hangover hit at age fifteen, the morning after a messy, hysterical hard cider binge to celebrate the end of a high school drama production—the bulbous, green 2-liter balloons of White Lightning supplied by our twenty-three-year-old professor (the one who also liked to tell his students how he got his inspiration drinking liquid opium). I vomited in the bathroom of my friend Bethia's mom the next day before traipsing, head thrumming, back to school, feeling edgy and exotic, like I'd stumbled into the plot of a Hollywood coming-of-age movie. My role? The quiet-yet-complicated one.

Which is where my fledging initiation into the grown-up "delights" of drinking falters. There was a new high in the hood for starters. The Ecstasy-fueled 1990s rave scene was already in full swing—half a crumbling DIY tablet washed down with a handful of water in the ladies' loos at the Camden Palace slaying inhibitions with a side of universal cosmic love. And the way it made the music *feel*. Oh. My. Goddess. Even better, there was no hangover. Only the yawning heartache of a lost love. Alcohol seemed so *basic* by comparison.

And then along came my first boyfriend, whom I shall call the Capricorn. A man six years my senior, to whom booze represented the root of all evil, having had his jaw broken in a drunken brawl. But who also smoked industrial-strength skunk weed from dawn to dusk. Forget wake-and-bake. This dude woke and then dedicated his entire day, every day, to getting as stoned as humanly possible. Alcohol would only . . . confuse matters.

Not to mention, it would perhaps incite the rebel in me. For the six years that we lived together, including the aforementioned college years, I was barred from imbibing. The Capricorn's was the kind of love that demanded total dedication, and my having any kind of a life outside the fortress of his devotion—especially the kind of life that might involve cocktails—was punishable by extreme emotional blackmail.

Cannabis, on the other hand, kept me quiet and meek, locked in my own private padded cell. Where soon I also chose to subsist on prison rations. An apple here, a handful of crackers there. Endless mugs of milky coffee. Shrinking my body to 30 pounds underweight created the illusion that I was still the one in control.

Years later, for my job as features editor for the UK's *Sunday Times Style* magazine, I wound up interviewing a friend of the Capricorn's sister. She'd gone on to become the first female president of a record label in New York City. Sitting across from her in the fancy penthouse restaurant of the company's building on Fifth Avenue, when she realized who I was, she practically spat out her *poussin*: "Oh my GOD. You're the *ghost* who lived in his room."

And it was alcohol that brought me back to the land of the living.

I was twenty-two when I graduated college at the top of my class. Surprise! I may have been a borderline anorexic with a daily weed habit, but still, mine was not one of those college educations that gets "pissed up the wall" (to use that quaint British expression) over happy hour doubles in the student bar. Feeling damn proud of myself, it was clearly time to celebrate what felt like a *big win* over what had been some *pretty shitty* goings-on behind the scenes. And we all know that any celebration worth its salt comes laced with liquor.

A Stella with my classmates here, a Chardonnay with the girls (girls who were already deeply suspicious of the Capricorn) there. Buoyed by my triumph, I began to apply a steady drip, drip, drip of booze to the vise-grip that had been my "adult" life to date. This was 1998, after all. The era of the "ladette," and the year *Sex and the City* sashayed onto our TV screens. Getting loose (read *freeing myself*) with pints of beer and hot pink cosmos felt like my birthright as a modern, emancipated woman.

And it turns out the Capricorn had been right all along. It was the summer of '98 that a combo of girl-powered bravado and sticky Sambuca shots danced me into bed with a tall, dark stranger. A calculated if cowardly move on my part, since I knew that Capricorn pride could be a prickly thing. Unable to muster the courage to leave him, he'd have no choice but to kick *me* out when he discovered this. All I had to do was fill the tanks with "lady petrol" and put my foot on the gas. And before I knew it, I was hurtling toward freedom—and what would become the rest of my life as I knew it.

• • •

Pause.

So here you have the beginnings of my personal *boozestory.* Nothing that out of the ordinary. Nothing to suggest the makings of a "problem" drinker (apart, perhaps, from the problem eating). But mostly it's just little old middle-class me, coming of age in a world where choosing your own adventure sometimes finds you face-to-face with the big bad wolf. In my case, a defining relationship that stripped me of my self-esteem. Robbed me of my voice. A world where, thank fuck, we've been taught that our moods are malleable and are presented with a full menu of options when it comes to hiding from our feelings, and checking out when the going gets *uncomfortable.*

And perhaps you saw snippets from your own boozestory too, mixed in with mine like cloudy pink bitters in a G&T—the allure of altered states of consciousness; booze as the bridge to full-fledged adulthood; cocktails as confidence in a cup; alcohol as a fast track to adventure. Perhaps you'll also concede a nod to our collective conditioning around alcohol being acceptable (not really a "drug," even) while other substances (MDMA, cannabis) remain taboo.

What qualifies me to talk about getting Sober Curious? The fact that my boozestory is *our* boozestory. Whether you learned to drink at home, at college, for fun, or as a way to numb your pain, you likely did not learn to question the ways in which we are taught to drink and think about drink. There's the occasional study on either the dangers or the benefits of booze. But unless your drinking is making you a danger to yourself or others, there is no rigid rule book when it comes to the right or wrong way to imbibe. Which I believe means each and every one of us is equally qualified to work this out for ourselves.

Because what if your experience of alcohol addiction (if you could even call it that) is awash with shades of grey? What if, from the outside, you don't appear to have a "problem"? You drink the same as, or even less than, your friends—can go weeks, months without craving or taking a drink. What if, as in my case, there is no violent, urgent, devastating reason to question your drinking? No rock-bottom. Only the hangovers, the occasional blackout, and the creeping suspicion that alcohol may be a, if not the, chief contributing factor to the overall sense of anxiety, ennui, and existential dis-ease clouding your days?

What if you find yourself at an Alcoholics Anonymous meeting anyway (as I eventually did as part of my Sober Curious journey), and because it feels like you're lying to yourself and everybody else, you choke on the word "alcoholic"? What if you cannot admit (as required in the first of the infamous 12 steps) that you are "powerless" over alcohol—and that your life has become "unmanageable" as a result? Because you're not, and it hasn't.

What if you're somebody who simply prefers not to drink—a preference that often finds you seemingly on the outside looking in? What then?

Well, it can feel like you're pretty much on your own, is what. Or are you? What if many more people than are ever likely to admit it (perhaps for fear of having to answer the Big Question for themselves: *Does the fact I'm asking all these questions make me . . . an alcoholic?*) are Sober Curious too?

It may surprise you to learn (as it did me) that the latest figures suggest that up to *one in eight* Americans is dependent on alcohol. *Millions* more of us than will ever find our way to AA or are even likely to think about our drinking that way. In my opinion, and based on my own Sober Curious experiences, this suggests that it's time to open up the conversation. To invite in some new boozestories, investigate some less well-trodden paths.

This was essentially the thinking behind Club SÖDA NYC (Sober Or Debating Abstinence), an event series featuring talks, meditation, guest speakers, and interactive elements that I launched in February 2016 (not to be confused with Club Soda UK, another sobriety support group based in London using the same, frankly, *genius* name). One thing I got super curious about when I went to AA was the role and value of community when seeking answers to complicated questions like these.

Since the only way to find my own Sober Curious allies was to begin asking some of my questions out loud, my idea was to create a space to talk openly about our often-conflicted relationship with this thing called booze—not to mention find some more answers for myself. Answers to questions such as: *How can we socialize and make connections without booze? Why can not drinking feel so awkward? How much is "too much," anyway?*

And another big one: *Why are the only people who don't drink the ones who* can't *drink?* Meaning, why don't more people see quitting drinking as a positive, healthy, life-affirming *choice*? Why do we often assume a person must have had a "problem" with alcohol when he or she quits? And why do so few people talk about the clarity, the self-assurance, the *presence* that becomes your very own inner lighthouse when you remove the booze?

• • •

So now that I'm getting this all down on paper, I am able to pinpoint the moment when I first got clear about my Sober Curiosity. There was this one Monday in the fall of 2010 when I clearly remember asking myself The Question:

Would my life be better without alcohol?

I was fresh off my first-ever yoga retreat, in Ibiza no less, the party capital of Europe, and something felt very, very different.

No sack-of-rotten-potatoes feelings were trailing me into the office. Rather, I felt optimistic and excited about the week ahead for the first time in . . . forever. It was also the first weekend I hadn't touched a drop of alcohol since . . . I couldn't remember. This felt like more than a coincidence. This felt like cold, hard evidence that my shitty Monday moods were linked directly to my weekend activity of choice: getting wasted.

Uh, duh!

Please don't roll your eyes. After fleeing the Capricorn I'd embarked on a full-on rebound relationship with booze that never really ended. Alcohol was so ingrained in my life by that time that I honestly hadn't ever considered that it might be a, if not the, major contributing factor to many of the problems I found myself grappling with. That my drinking might be the very thing slowly and stealthily decimating the overall quality of my life.

There was the daily tooth-grinding, gut-churning anxiety that had become my almost constant companion, and the overall ennui that descended when I wasn't in the grips of said anxiety. There was the waking up most nights at around 3 a.m.—heart pounding in my chest, mind feeling like feeding time for the monkeys at the zoo. There was my nonrelationship with my mother. My persistent IBS. And coloring it all, the sense that something *dreadful* was lurking around every corner, waiting to wreck my reputation, my career, my marriage, my life.

Again, I share all this because I have a feeling my "symptoms" may be familiar to many of you—the anxiety and hopeless *lack of meaning* I drank to forget, and that I have since learned are *also* often the symptoms of a run-of-the-mill, not-nearly-as-uncommon-as-we-think, dependence on alcohol. In the brilliant *This Naked Mind: Control Alcohol* (2015) by Annie Grace, something of a Sober Curious bible, she suggests this: "We continue drinking to get rid of the empty, uneasy feeling that alcohol created [after it left our system]. When we enjoy the 'pleasure' of a drink, we restore the wholeness and peace of mind we knew our entire lives before we drank a drop."

I can't say that it was as simple as that for me. I had not, at that point, ever in my adult life experienced true wholeness and peace of mind, having skipped over, on a raft fashioned from beer bottles and cocktail umbrellas, the part where I truly healed from my eating disorder and my relationship with the Capricorn. But this did give rise to another question: *Could it really be that the anxiety, the ennui, and the sense of dread were* equally *the by-products of my garden-variety* cravings?

This is one of many paradoxical elements to this story: like how we drink to feel more confident and often wake up in a sea of self-doubt; like believing it's "cool" to drink when often we do it to fit in; like drinking to make us feel sexier while instead it numbs our senses and makes us look like shit.

Another thing I quickly discovered was that navigating these paradoxes and answering all my Sober Curious queries for myself was not going to be easy. Following my experience on the yoga retreat, did I abstain from boozing the following weekend? I did not. Life without Friday night drinks was . . . *unthinkable.* But what I did do was embark on successive failed attempts to moderate my intake. Never during the week. Okay, never more than *two glasses.* Alright, yes, I'll have a top-up. You know how it goes.

Don't you?

I have also developed a little theory along the way, which I'll introduce you to in chapter 1, which is that anybody who drinks on a regular basis is probably, kind of, just a little bit addicted. Which is not supposed to scare you. Or come across as judgmental. But a big part of my Sober Curious path has been about dismantling the stigma that tends to color our perception of addiction. The reason I had such a hard time "joining in" when I found my way to AA? "Ruby, alcoholic" did not and does not feel like my story. It felt like a life sentence, laden with shame.

Being "Ruby, Sober Curious," however, has most definitely meant acknowledging that I *have been* addicted to alcohol. This, in turn, has meant extended periods of total abstinence, since getting unaddicted to booze has one basic requirement: *not drinking it!* It's also meant feeling a fuckload of feelings, facing untold awkward "Sober Firsts" (more on these later, too), making new friends, losing some old ones, and finding new ways to feel free and let loose.

As far as life adventures go, getting Sober Curious has been as fraught with challenges as it has been unexpectedly joyful and suffused with bliss. It has helped heal parts of me I didn't even know were hurting. It has brought me the clarity I needed to confront the demons of my past once and for all, the self-assurance that I am the one who's in control of my life, and the presence of mind, body, and soul to take action on my biggest, boldest dreams and visions. In short, it has been both the undoing and the making of me.

• • •

So, what might it mean for *you* to get Sober Curious?

Again, this is not for me to predict. Your journey must and will be yours and yours alone. What I can promise is more energy, vitality, and confidence in your life choices than you may have ever experienced. Here's an overview of what it means for me.

First and foremost, being "Ruby, Sober Curious" means thinking of myself as a sober person: somebody who does not drink alcohol. And it also means not having any rules about not drinking—since we all know rules are made to be broken. It means not trying to make any rules for anybody else about not drinking, too. And, for me, it means leaving space for an immaculately informed and superconscious choice to take a drink on occasion. (For anybody whose hackles immediately begin to twitch at this—*"but drinking 'on occasion' makes you NOT a sober person!!"*—we'll dive into these absolutely valid concerns in chapter 1.)

In my day-to-day life, it means often being the odd-one-out . . . and my being very much okay with this. It means not making up excuses about why I'm not drinking to make me or anyone else feel more comfortable. It means early nights, orgasmic sleep (more on this later too!), and every morning feeling like a fresh start. Most of all, it means being honest with myself about the way alcohol makes my mind, my body, and my *soul* feel—and trusting the *wisdom of my being* rather than the external messaging about alcohol we are bombarded with daily.

As for how to navigate this unconventional, *controversial* lifestyle choice? Well, the clue is in the title. Ultimately, living Sober Curious has meant getting *curious* and seeking to answer with as much integrity as I can muster all the questions about alcohol with which I've peppered this introduction. To recap, some of the bigger ones for me have been: *Would my life be better without booze? What is 'normal drinking' anyway? Why is alcohol so . . . everywhere? What would it be like to* never *wake up with a hangover?*

The aim of this book is to help you find your own answers to questions like these, and more—questions you will begin to formulate only as you step fully onto your own Sober Curious path.

Last but not least, my personal Sober Curiosity has also been the surprise subplot to the spiritual awakening I wrote about in my first book, *Material Girl, Mystical World: The Now Age Guide to a High-Vibe Life*—a spiritual awakening the journalist in me can see unfurling all around as we collectively begin to question what truly brings meaning to our lives; as we seek the clarity to make life choices that are aligned with our individual needs—with faith that the uncertainties we're faced with are in service of a bigger picture and the courage required to stay present in a world so full of distraction—so that we might contribute something of value.

Not to mention that connecting to my spirit has shone a beam of truth on my personal boozestory—namely, that alcoholic *spirits* have been a second-rate stand-in for the joy, inspiration, confidence, connection, and overall sense of *aliveness* that I now know I am perfectly equipped and cosmically designed to generate for myself.

It's a journey shared by many leading voices in the modern self-help space, such as Brené Brown, Glennon Doyle Melton, Russell Brand, and Gabrielle Bernstein—all of whom have been open about their struggles with addiction. As the alcohol fog has lifted, the wider story of our collective (and often secret) love-hate relationship with booze has come into laser-sharp relief. Not only do I believe that alcohol is keeping millions of us from living as our happiest, healthiest selves, but one of the common "side effects" of getting Sober Curious (or just Sober Sober) is a strong desire to help others. Meaning that the possibilities when more of us start living "hangover free" are endless.

So, thank you for being brave enough to get Sober Curious with me. For choosing to trust your body, to go against the grain, and to get into a staring match with stigma.

On the part of your Sober Curious path that we'll be walking together, I'll be sharing all the tools you need to feel supported, confident, and informed; to make educated, fully conscious choices about alcohol, and in doing so to realize your full potential as a co-creator of our collective future thriving.

So what d'you say—are you in?

1

THE NATURE OF THE BEAST

NEW YORK CITY, FEBRUARY 2015

Cut to my first AA meeting: dry mouth, pounding heart, as the infamous roll call makes its way around the room. "Adele, alcoholic," "Mike, alcoholic," "Susan, alcoholic." And then it's my turn. "Hi . . . I'm Ruby and I . . . think I . . . *might* be an alcoholic?" Another question. But nobody answers, and the list continues. "Jim, alcoholic," "Leslie, alcoholic." (Names have been changed.)

The meeting is taking place in a church basement in the West Village in New York City, and given the associations I had with the word "alcoholic," people are much better dressed and have much better hair than I expected. Also, they're super articulate. It's the kind of meeting they call a Big Book meeting, which means somebody reads a passage from the AA bible (the Big Book), and then other people share on the themes. The vibe is vulnerable and raw, cradled by the wisdom of elders. I am here to listen, to look, and to learn. Because, despite having drastically reduced my own boozing by now, I still find myself doing plenty of *thinking* about drinking.

Being a journalist, I've also been doing my research, and according to the American Society of Addiction Medicine, "preoccupation with substance use" is one of the criteria for any substance use disorder, whether you're using said substance compulsively or not. And so here I am, having decided it's time to answer the Big Question—*Does the fact I'm asking all these questions make me . . . an alcoholic?*—for myself.

And, because I'm nosy. Maybe it's also the journalist in me, but I've always, even before the questions began, wondered what goes on behind the closed doors of Alcoholics Anonymous. A secret society that's open to anyone—at a price: not one more drink, for the rest of your life. At least, that's my perception. Though AA preaches total, lifelong abstinence as the only "cure" for addiction to alcohol, plenty of people who attend AA regularly relapse but still find support in the groups.

For somebody like me—a regular ol' *social drinker*; the one they used to call "Cocktail Girl" on holiday because of the fab concoctions I conjured from the cheapest Cava money could buy; a girl who only *ever* used alcohol to max out the fun (question: *Didn't I?)* and who barely drinks now anyway—nostalgia for the "good ol' days" means I'm reluctant to commit. Aren't there likely to be future situations where drinking could still be fun? A glass of champagne to toast my best friend's wedding? A beer in the sun to loosen my hips at an open-air concert?

Even if at some point alcohol appears to have morphed from my leisure activity of choice to presenting something of a problem (if not the kind of problem that's likely to land me in jail, the ER, or the morgue), the thought of *total, lifelong abstinence* being the only solution kind of feels like being led to the edge of a cliff and told that down there, if I can only muster the courage to leap, I'll find nirvana. Just close your eyes, hold your breath, and jump.

Nope. Never gonna happen.

Maybe We're All a Little Bit Addicted

So how (I found myself asking, as I texted the super successful, articulate, and stylish AA friend I was finally asking to take me to the meeting) did I wind up here?

Jail, ER, the morgue—just a few of the places some people in AA will tell you they are likely to wind up if they break the vow of total, lifelong abstinence. Where one "innocent" drink could well land a seasoned alcoholic. Which fits with the image we often hold about somebody with a drinking problem. Perhaps it means waking up with the shakes and needing a drink before work (if a job is still in the picture). Maybe there's a trail of wrecked relationships and a stack of DUIs. A chronic health problem. None of which accurately describes the kind of drinking I was doing by the time I went to AA. Or have ever done.

The alcohol problem I have since identified in myself was that booze was preventing me from being *fully present in my life*. By which I mean, it was preventing me from knowing, in each and every moment of each and every day, what it really felt like to be ME—a "problem," since it's only from this place of knowing, of presence, that I can truly choose which decisions to make and which next actions to take in service of my own highest good—in order to create a life that feels *meaningful* to me.

Which was a beast of a different complexion entirely, wasn't it? One that would require, it seemed to me, a different kind of approach. My getting comfortable with being present, for example. Seeking tools and practices and philosophies to help with this. From there, working out exactly what *did* hold meaning for me. And then mustering the courage to make the necessary changes and take the necessary risks to create a life that looked like that.

When "Ruby, alcoholic" stuck in my throat, it was because it felt like a lie. Like strapping myself to a part of my life that was already fading fast, like a neon road sign being left behind in the dust. The chariot I was by now riding, full-throttle, into my more present and meaningful new life, was being fueled by the mystical and emotional healing tools I was writing about in *Material Girl, Mystical World*. But more on what these looked like elsewhere, because first, I was still faced with removing the dregs of this substance from my life. I might have radically shifted my drinking habits, but the well-worn grooves of my Cocktail Girl persona ran deep.

Which brings me to the conclusion I've arrived at as a result of my Sober Curiosity: that as humans, *we're pretty much hardwired to get hooked on hooch,* to find ourselves more dependent than we ever anticipated on our beloved booze. If "Ruby, alcoholic" did not feel like my truth, then, as noted, I will happily acknowledge that I *have been* addicted to alcohol, and that I had developed what I might call *deeply ingrained habitual drinking patterns*—just like so many more of us than would ever identify with the term "alcoholic."

I mentioned in the intro that my theory goes as far as to suggest that anybody who drinks on a regular basis is probably, kind of, just a little bit addicted. Here's how I think this works.

For one, as opiate-addict-turned-brain-scientist Marc Lewis writes in *The Biology of Desire: Why Addiction Is Not a Disease* (2015), our brains are designed to seek out and repeat any experience that either brings us pleasure or helps us avoid our pain. It's an evolutionary tic as old as time that's designed to keep us out of harm's way as we go boldly forth and procreate, while trying not to die before our time. Alcohol, fleetingly, and on a very superficial level, appears to do both these things.

For two, booze is heavily marketed to us from the age that we're old enough to understand. Consider this: Drinks company Pernod Ricard alone invested $421 million in advertising in the United States in 2017. But booze is also sold to us by watching how our friends and families use drink. By the way they do it in the movies. By chalkboards outside bars advertising the happiest hours of the day. And by our own innate desire to transcend the daily trials and traumas of being human. The way we *learn* to see it, alcohol is pretty much the elixir of a life worth living. A life that involves laughter, connection, relaxation, and inspiration, that is.

And then, for three, according to a 2007 report published in the medical journal *The Lancet*, alcohol also happens to be one of the five most addictive substances on the planet—up there with heroin, cocaine, barbiturates, and nicotine.

Since this realization—that is, anybody who drinks on a regular basis is probably, kind of, just a little bit addicted—has been pretty integral to my overall Sober Curious awakening, I'll go ahead and repeat:

1. **Our brains are biologically hardwired to form an attachment to alcohol.**
2. **Alcohol is universally presented as both good times in a bottle and a panacea for a multitude of modern malaises, as we are actively and regularly reminded.**
3. **Alcohol is *as* addictive, if not *more so*, than cocaine.**

The reason this was such an important breakthrough for me? Acknowledging and accepting that I was probably, kind of, just a little bit addicted to booze gave me all the insight I needed as to how to remove its hook from my psyche once and for all. Explained why my attempts to "moderate" my intake (*never during the week; well, never more than two glasses; okay, I'll have a top-up*) had never worked. If not suffering from the chronic, progressive brain disease of alcoholism, as a garden-variety *habitual drinker*, I knew what I had to do—which was to begin thinking of myself as a sober person. Somebody who does not drink alcohol, period. While also embracing the possibility that this may or may not look like total, lifelong abstinence for me.

This may be a little confusing, given the black-and-white approach of the AA/12-step model that we're taught is the "cure" for addiction. But increasingly, research shows shades of grey when it comes to addiction. Did this mean there could be shades of grey in sobriety, too? This is essentially the philosophy that has formed the basis of my own Sober Curiosity, and that we'll unpack in more detail in this chapter.

Which is not me trying to go ahead and answer the Big Question on *your* behalf. You may well have never even considered the words *alcoholic* and *addiction* in relation to your drinking. I simply share this observation to illustrate that when it comes to the duel of us versus booze, the odds are most definitely *not* stacked in our favor.

Which also, for me, helped remove some of the stigma and shame I'd been wading through—stigma and shame, as we will discover, that also propagate addictive behaviors of all flavors. Perhaps, given the nature of the beast, my continually seeking out and consuming booze did not mean there was something *wrong* with me, did not mean I was "diseased." Perhaps it just meant I was . . . *a human being!*

Time for another question: *If my story could be many of our stories, then when does "normal" drinking end and Sober Curiosity begin?*

How the Habit Takes Hold

Something else I've learned is that journalism (within the field of "arts and entertainment") is a high-risk career for alcohol abuse. In case you're *curious*, only hospitality and construction work see higher rates of substance use disorders. When it comes to my chosen profession, I blame the stress—of deadlines, of competing for stories and readers, and of putting your name and opinions out there in a very visible way (like I am also doing here).

And then there's the endless merry-go-round of free drinks.

After graduating college, I walked straight out of the Capricorn's clutches and into my first magazine job. Based in a makeshift office above a bar in trendy SoHo and funded by a bunch of East End gangsters, it wasn't exactly the most professional setup. When the "lads" would swing by the office Fridays to dish out wraps of cocaine and crisp, pink £50 notes by way of payment for our services, I took the fifties and declined the coke. As it had been another evil on the Capricorn's banned list, I was deeply suspicious/scared of what was (and remains) fashionable London's drug of choice.

Alcohol, meanwhile, had swiftly become my trusted sidekick, as I navigated a new life "on the scene." Every night brought another product launch, venue opening, or artist showcase—all of it floating merrily down a darkly seductive river of free alcohol, a pleasure boat draped with fairy lights and fabulous new friends. And my name was on every list. For a "quiet yet complicated" girl like me, a former "ghost" of a person, it was like somebody had turned the color on. Which is when the warning light also begins to blink, a flicker of pale amber hinting at the possibility of danger ahead.

As Marc Lewis writes in *The Biology of Desire*: "When our experience of the world is fraught with strong feelings—whether of attraction, threat, pleasure, or relief—brain changes take on extra momentum . . . Most specific to addiction, the feeling of *desire* for something specific shapes the brain more acutely than other feelings."

The "brain changes" he's talking about are the new neural pathways that form when we repeat any experience that the brain likes, meaning an experience that either brings us pleasure or helps us to avoid pain. The deeper these neural pathways become, the more likely we are to unconsciously repeat the behaviors that led to them—meaning repeat them *without question*. These are also the brain changes that have led the medical establishment to label addiction a "disease," a diagnosis we'll get into and that Lewis has set out to debunk with his work.

But back to that amber warning signal. Was I having some very strong positive feelings about my newly minted role as in-demand girl-about-town? Hell yes. Did I *desire* lots more feeling popular and cool, versus feeling like a ghost in my own life? I most certainly did. And was I also teaching my brain that imbibing copious amounts of alcohol was an integral part of my experiencing these new feelings? Abso-fucking-lutely.

You can apply the above equation to your own boozestory by simply filling in the gaps: *"Was I having some very strong feelings about X? Did I desire a lot more feeling X, and a lot less feeling Y?"*

In my case, you can also throw in a new relationship, with a man I will call the Pisces, who I knew from the moment we first shared air space would be the love of my life. Even better, the person I now *desired* with the entirety of my mind, body, and soul (and to whom I remain happily married twenty years down the line) was a DJ and a party promoter. When he landed a gig launching hot new nightclub Fabric, it came with a card for unlimited free drinks in the VIP area (where I wound up proposing to him one leap year—yes, drunk). Now, my nights off were spent dancing on speakers, my Sundays "recovering" over pints of lager in sunny, soft-focus pub gardens.

As I alluded to in the introduction, everybody in my world drank as much, if not more, than I did. And as Amy Dresner, author of addiction memoir *My Fair Junkie: A Memoir of Getting Dirty and Staying Clean* (2017), notes in a Q&A that went out with the press release for her book: "When you're using drugs daily, you're usually surrounded by other daily drug users. In that setting, it's easy not to see yourself as an 'addict.'" Especially when your drug of choice is *universally presented as both good times in a bottle and a panacea for a multitude of modern malaises.*

And the years rolled by. The publications I worked for became more respectable, my roles on them more demanding, and my relationship with the Pisces as cozily domestic as they come. He got a "normal" job in PR, and in the name of "balance" I began to actively stake out several nights a week when I wouldn't drink at all.

When I landed my dream job at the *Sunday Times*, I took an extended break from weeknight drinking altogether. I wanted all my wits about me. (Question*: Why wouldn't I always?*) But since booze was fully confirmed by now as my fast track to relaxation, to joy, to adventure, and to connecting with the people I most loved and who loved me, why would I have ever contemplated letting it go completely?

No More Moderation or Pretending

Before we go any farther down this rocky part of the path, let me get clear on something. Rule number one of changing your drinking habits is: *You have to change your drinking habits.*

I repeat. If you're slowly getting on board with the idea that anybody who drinks regularly might be kind-of-just-a-little-bit-addicted-to-booze, then the only way to get unaddicted is to STOP DRINKING BOOZE. Same goes for if you just want to drink differently from the way you have been drinking. Or even want to *think* differently about drinking. After all, in the words of Albert Einstein: "No problem can be solved by the same kind of thinking that created it."

This may or may not look like *total, lifelong abstinence* to you, and this is perfectly okay. But this is also not about "moderation."

As noted, by the time I went to AA I was already drinking considerably less—had cut it back to once or twice a month and to well within those government guidelines of "seven units" per week. Even if the thinking about the drinking carried on relentlessly.

After my first meeting I went back once more, just to be sure, and came to the same conclusion—that "Ruby, alcoholic" was not my truth. And so, I simply carried on as I had for the previous five years—which is to say, slowly unpicking my *deeply ingrained habitual drinking patterns* one dry dinner date, euphoric Sunday morning, or guilt-ridden hangover at a time. Walking my own (albeit imperfect) path and *choosing* not to use alcohol on my own terms had thus far proved overall to be so positive and life affirming. Acknowledging my addiction to alcohol also helped me feel more secure in this choice. And I wanted more of that.

Let's not forget that this "choice" is also a privilege, as my friend Holly Whitaker, a badass feminist sober coach and founder of AA alternative Tempest, likes to remind me. Meaning I am fortunate that my drinking never got so bad that I *had* to stop. And speaking of privilege, let's also thank Goddess that an organization like AA—which offers unlimited, *free* support for those struggling with addiction—even exists.

Never having hit rock-bottom with my drinking I'm also what Holly calls an "early-exiter"—and it's interesting to note that she also believes that "people who have struggled less [with alcohol] have a harder time making the hard break." Having chosen total, lifelong abstinence for herself, she sees alcohol as "a toxic substance that, by the way, causes weight gain and breaks your capillaries. And makes you do stupid things. That is actually a depressant, and also breeds anxiety. And makes for really bad, shitty sex. I mean honestly, is the purpose of human existence really to try and find a way to keep this *in* our lives?" With talk of "moderation," she means, versus just *not drinking.*

But she also takes a more lenient stance on abstinence being the *only* path to changing your drinking. "Your integrity is what matters, and your trust in yourself," she told me. "On the one hand, we [at Tempest] hold this absolute idea that alcohol serves no purpose. Along with the kindness and compassion for people to come to that conclusion through their own experiences, and on their own terms."

The privilege part really kicks in with the awareness that as a wealthy, educated, white woman, the world has also led me to believe that my opinion about my life counts. That it is within my power to get educated and decide for myself what's best for me.

All of which is *not* to say that I'm anti-AA! Or anti–total, lifelong abstinence, if that feels right for you. Another thing about the AA meetings that felt super empowering was the support of being *in community*—a community where people have a space to share their boozestories without fear of judgment. You could feel the collective outbreath as people, liberated from the *prison of pretending*, filed into the room, like members of a Broadway chorus line shrugging off costumes a few sizes too small. Ahhh . . . the relief of no longer having to fake being "normal"!

Well, I wanted more of that, too! Because what's one thing that fuels obsessive behaviors, including obsessive thinking? This one I knew from my brush with anorexia: *doing it in secret.* Which is about when I decided to create my own kind-of-just-a-little-bit-addicted-to-booze support group.

The first Club SÖDA NYC meetup took place in my living room and was composed of a handful of women I suspected might be feeling the same about booze as I was. How did I know who to invite? Think about the people in your life. Whether it's the friend who always drinks until she blacks out, the one who doesn't drink because he's "allergic," or the colleague who tends to suggest a yoga class or softball game instead of after-work drinks—you just kind of *know*.

Of the fifteen or so people I emailed, about seven showed up. We had snacks and sat in a circle and simply shared, in a very AA kind of a way, the impact that alcohol had on our lives. The obsession. The stigma. The regretful texts. The terrible sex. The way it promised *more* confidence, and then seemed to rob us of it. How *not* drinking made us feel amazing . . . and also like weirdos, outsiders, the most boring bitches in the office. What made us even want wine in the first place.

And it was my most liberating Sober Curious experiment yet.

Whether coming together like this meant we were all alcoholics and now had to stop drinking completely? Well that was *another* question.

Changing a Habit Is the Hardest

Often when I talk about my previously *deeply ingrained habitual drinking patterns,* I follow it up with this story.

When I was a kid, my mom was really involved with an Ayurvedic doctor/guru named Shyam Singha, who believed that you could cure any and all ailments with food. Pretty progressive in Britain in the 1980s. His theory stretched to addictive behaviors, I guess, since the one Shyam'ism that's always stayed with me is this: A habit is very hard to get rid of. Take away the "H" and you still have "a bit." Take away the "A" and you still have "bit." Take away the "B" and you still have "it"!

He obviously didn't see alcohol as particularly problematic, however, since he also had a diet called "Weight Loss Plan for Rich People": three filet steaks and one bottle of champagne per day.

But I digress.

Habitual behaviors are those we perform unconsciously, without thinking. *Without question.* Our brain is so used to performing them, it knows it can save us time and energy by having us do them on autopilot. So, when does a habit become an addiction? One definition is when you become *conscious* that you're habitually doing something you do not want to do, and you keep doing it anyway. Which is very often the case with habitual drinking.

The psychology of this cycle creates what's called "cognitive dissonance," which is what happens when we attempt to reconcile two attitudes or beliefs that contradict one another (we know that drinking is bad for us, but we do it anyway). The internal push-pull and futile struggle for "cognitive consistency" leads to a constant feeling of fighting against one's own nature. Jeez, it makes me anxious just writing about it.

Want to simplify things? Remember. *Rule number one of changing your drinking habits is . . . you have to change your drinking habits.*

Alcohol addiction, or the habit of drinking, becomes a *dis-ease,* meanwhile, when the neural pathways associated with the behavior of drinking are so deeply ingrained they have changed the way the brain functions. This often means the "diseased" individual will go to great lengths to perform said behaviors, risking personal safety and interpersonal relationships, for example.

Alcoholism was defined this way by the American Medical Association in 1956, and for many it's a diagnosis that helps keep things black and white (answering our human need for structure, and in doing so keeping millions out of jail, the ER, or the morgue). Not to mention, as Marc Lewis writes in *The Biology of Desire,* that "disease" is also "how medicine defines human problems"—often because "it's many times easier to perform impressive research on the workings of cells than the workings of families or cities—because that's where the money is, and because cells are easier to observe."

Which begs another question: *What if the remedy for alcoholism, and/or addiction, actually lies in researching, meaning getting "curious" about, the ways in which we live, love, and relate to one another?*

I was introduced to Lewis's work by one of the women at that first Club SÖDA NYC meetup, and his book was a *game changer* for me. It was where I first learned how our brains work when it comes to seeking pleasure and avoiding pain. Leading to my theory that addictive drinking is actually fairly *standard.*

Given the prevalence and normalization of alcohol consumption, could it be possible that it's harder *not* to become alcohol dependent, to develop the habit of drinking, than it is to maintain a perfectly "healthy" relationship with booze?

When I shared my theory with Lewis in an interview for this book, he went so far as to add that "'normal' drinking might well be addicted drinking. There's no fundamental reason to think that addiction isn't normative."

As noted, for me, thinking about addiction this way means there's instantly less *shame* in admitting that *you too* have less control over alcohol than you'd like to think. The way to change a habit? Wake it from its seductive slumber by sounding the fire alarm and shining the megawatt light of awareness directly in its face. Ugh, so *irritating.* You were sleeping! But keep pressing snooze and you will surely be engulfed by flames.

When it comes to learning to trust in *your own power over your own life*, more awareness—meaning less pretending, less hiding, less *shame*—is always the aim.

In one of her earliest works, *I Thought It Was Just Me (But It Isn't): Making the Journey from "What Will People Think?" to "I Am Enough"* (2007), researcher (and openly sober person) Brené

Brown writes that "addiction and shame are inextricably connected. They are also very similar: Both leave us feeling disconnected and powerless . . . Addiction can make us feel alone and on the outside [*because there is something wrong with me*—the internal messaging of shame] . . . [and] there is often a sense of secrecy and silence about addiction."

For me, thinking about addiction as a disease, for which abstinence is the only cure, was only fuel for the cycle of shame that kept me locked in repetitive unhealthy patterns with drinking. It goes something like this:

> Shame stems from the belief that "there is something wrong with me."
>
> Alcohol provides the temporary illusion of belonging and relief from feelings of shame.
>
> Because of alcohol's addictive nature, using it regularly inevitably leads to alcohol dependence.
>
> Chronic alcohol dependence is termed a "disease."
>
> Identifying myself as "diseased" means "there is something wrong with me."
>
> I feel a new kind of shame, also tinged with the fear that this disease is eventually going to kill me.
>
> Alcohol provides temporary relief from these new feelings of fear-tinged shame. I become further dependent on alcohol. I am more deeply diseased. I feel more shame.
>
> And so on.

Ugh!

It's also worth noting here that being "shame-prone" is more common in those who perceive themselves as being "other," defined by Brown as women, people of color, sick people, poor people, divorced people, children of divorced people, overweight people, people who have experienced sexual abuse, people who have experienced bullying . . . the list goes on.

In other words, these are often the "some people" who might not have been led to believe in *their own power over their own lives.* Might not believe that they too deserve to have their own narratives, to make their own choices, and to form their own opinions as to how their lives play out.

Recognize yourself and anyone you know in Brown's list? Honestly, it would be "other" if you didn't.

Turn Every "Relapse" into a Reminder

If you've digested the above, it probably shouldn't surprise you to learn that between 40 and 60 percent of people in abstinence-based recovery programs relapse (meaning begin using again). Marc Lewis has a reason for this, and it's called "ego fatigue"—when "self-control begins to blink and fizzle like a light bulb." It's essentially the end of willpower.

Since ego fatigue is pretty much inevitable in humans and tends to happen even faster when it comes to tasks "that require you to suppress your desires, hide your emotions, or ignore important information," I prefer to term my relapses "reminders." It just feels kinder, doesn't it? Less shameful.

In my experience, each time I drank again, having vowed not to, once the hangover and the initial self-flagellation had subsided my resolve was only *strengthened*—leading to ever-longer periods of easier-to-sustain abstinence. And the longer I abstained, the more unexpectedly joyful I found sobriety to be, the less I felt any desire to drink, and thus the "reminders" became fewer and farther between.

The above cycle also reflects a shift from my feeling *shame* to my feeling *guilt* about my lapses. Which is a good thing! According to Brené Brown, it means I went from believing there was "something wrong with me" (like, I was diseased and powerless against alcohol) to believing I had "*done* something wrong"; that is, the relapse/reminder was something outside of myself that was in my power to *choose* to not repeat in the future. After all, as Brown puts it, "it is much easier to change or fix any defective behavior than it is to fix a defective self."

This also reminds me of a saying a friend of mine once coined: "Perfection is a prison, but excellence is elastic." Because the real reason I choose to be Sober Curious, and not just Sober Sober? For apparently going to such lengths to keep Holly's utterly pointless toxic poison in my life? Not because I still desire the momentary buzz I might get from it, or because I think my life is any better with alcohol still in it—but because I don't want to live with the pressure of trying to be perfect.

My innate perfectionism (stemming from the shame of my own "otherness": woman, divorced parents, "poor kid" in a wealthy school, etc.) was partly what fueled my teenage anorexia. And I think it's one thing that drew me to drink in the first place. I was surprised when I learned, way back at the beginning of this journey, that alcoholics often self-identify as perfectionists—since the image we're presented with is anything but.

We're taught to see addicts as weak, sloppy, self-indulgent, lazy. Which are also just parts of being human, by the way. What, you never drop the ball? You're never vulnerable? Confused? Selfish? You never can't be bothered? Trying not to be these things—to even *appear* not to be these things—can get pretty exhausting. In my case, alcohol gave me permission to forget trying to be perfect and to allow life to just be a *beautiful mess*. For Saturday night, at least—the price being the *ugly* mess inside my head the morning after.

But the years passed. And the farther I leaned into my Sober Curiosity, each time I drank became a mindful and fully conscious choice—one I often entered into as an experiment of sorts. An experiment designed to answer yet more questions: *How will this drink make me feel? Will it enhance the situation or detract from it? What are all the reasons—emotional, social, physical—I'm choosing to drink?*

These experiments resulted in a watershed moment last summer. On a flight to Europe traveling to present some workshops at an island festival in Croatia, I realized I hadn't once considered having a beer—only *after* the flight attendant finished serving drinks. Old me would have been debating this choice for several weeks in the lead-up, since drinking "on vacation" was one of the doors I'd always left open.

The old questions might have been: *Will I have just one? Or will one lead to three? Exactly how much will I hate myself if that ends up being the case?*

I honestly never thought I'd be the girl who forgot they served free alcohol on international flights. After all, I had learned that the only way to "recover" from addiction to alcohol is with *total, lifelong abstinence*. And the longest I'd been completely abstinent by that point was probably a few months.

Later that week, I hosted a sober New Moon celebration/rave in an open-air amphitheater at said festival. I hadn't billed it that way (as "sober," I mean) because I didn't think anybody would come if I did. But two hours in, the house music pumping, crowd gyrating as one ecstatic entity, people kept coming up to me yelling: "This is amazing! I can't believe I haven't had a drink!"

I couldn't believe I still hadn't, either. I was up on the stage for most of the night, shaking my ass like in the *good old days* with the Pisces, when I would have been fueled by unlimited VIP cocktails—but even more wildly! It was another breakthrough moment.

Walking home through the pine trees, my body vibrating all over and high as a kite on vibes, I found myself wondering: Did the fact I could even dance my ass off sober now, not having thought about nor missed having a drink, but still not having signed up for total, lifelong abstinence, mean I'd officially beaten the system?

Perhaps.

As for how much I drink these days? I'm not going to tell you. You only want to know so you can compare it to how much you drink and then use this as an excuse to either let yourself off the hook or beat yourself up some more. I know you too well! And anyway, how much *I* drink is irrelevant to *your* Sober Curious story. Remember what Holly said: *"Your integrity is what matters, and your trust in yourself."*

It could be said that living with integrity, meaning seeking to align your every thought, word, and deed, is what breeds cognitive *resonance*—that is, the sense of balance and calm that comes from aligning our thoughts and deeds, and the antidote to the turmoil of cognitive dissonance. All the spiritual gurus will also tell you that this is the fast track to inner peace, lasting confidence, and ultimate well-being.

I predict that by the time you've finished this book, you may feel the same way as Annie Grace: "I drink as much as I like, as often as I want." Since when you *truly* get to choose—having gotten Sober Curious and answered with as much integrity as you can muster every question about you and booze—you may well find that drinking as much and as often as you like means "none" and "never." That you might even choose total, lifelong abstinence.

2

CONQUERING FOMA (FEAR OF MISSING ALCOHOL)

As I found myself contemplating the sheer cliff face of total abstinence, the fears swirling in my frontal lobe (the part of the brain that monitors the potential outcomes of our actions, as well as suppresses socially unacceptable behaviors) went a little bit like this: life will become dull and monotonous; I'll lose all my friends; people will think I'm a weirdo, must have a "problem," or worse, that I'm all judgy and superior; it will be impossible to "switch off" from work; holidays, weddings, and birthdays will never be the same again; the Pisces and I will have nothing to talk about; I'll never have a reason to get dressed up in an *outfit*.

This last one seems so ridiculous, doesn't it? The idea that without cocktails, what would be the point of ever putting on a dress and a pair of heels? And yet it surfaced time and again. Proof of just how deeply ingrained alcohol was in my experience of fun times, parties, *celebration.* And, on a deeper level: of participating in the world as a person who felt comfortable being seen. After all, why put together a killer look unless you want to be looked at?

As for my other fears, I suspect you may be familiar with at least a few of them. Given its role as the ultimate social lubricant, bonding tool, and all around "awesome water," alcohol, for most of us, is intimately woven into the fabric of our social interactions. Whether we're at home, at work, or on a hot date, a glass of wine or a cold one is the universal code for "let's connect."

And we humans are "hardwired for connection," with study after study now showing that the quality of our relationships is by far the greatest predictor of overall happiness. Also that "we are profoundly shaped by our social environment and we suffer greatly when our social bonds are threatened or severed," says Matthew Leiberman, author of *Social: Why Our Brains Are Wired to Connect* (2013), adding that "social pain is real pain."

It's no wonder that the thought of removing booze from our Friday night activities leads to visions of our sitting alone glued to our phones or bingeing on bad TV, depressed and dressed in the same food-stained slouchies we'll be wearing all weekend.

Welcome to the world of FOMA: Fear of Missing Alcohol.

But there's good news. While walking the Sober Curious path has meant I've confronted a lot of these fears head-on (meaning, a lot of them have proven to be rooted in some pretty uncomfortable realities), I've also come to realize that *the only thing you miss out on by not drinking is . . . getting drunk.*

Mic drop!

My point? That far from being integral to our experience of connecting with others, and therefore living a full and happy life, getting drunk is an activity in and of itself: a simple act of altering our mental, emotional, and physical states through imbibing toxic liquor (a substance that, as Holly Whitaker reminded me, "is literally the same thing we use to fuel rockets and cars"). One that actually makes us *less* aware of our surroundings and the people we're with. And that, as many of us have experienced, can also just as easily lead to arguments, fights, texts you regret, terrible sex, and, in the cold, queasy light of a morning after, our feeling utterly, devastatingly alone.

Get to Know What Triggers Your FOMA

Not that we usually choose to remember it that way. Enter the phenomenon known as *euphoric recall*. It might sound like a 1990s Arnold Schwarzenegger movie, but euphoric recall is actually a psychological term that Wikipedia defines as "the tendency of people to remember past experiences in a positive light, while overlooking negative experiences associated with that event(s)." Euphoric recall is often cited as a factor in substance dependence, and I first heard the term from a friend in recovery and have used it since as a way to recognize when I'm getting all misty-eyed about my boozing—and therefore more likely to succumb to FOMA.

The reasons we drink, and the situations, people, places, and memories that will likely lead to bouts of FOMA and/or euphoric recall, are going to be different for everyone. As individual as our thumbprints—and another reason that a one-size-fits-all approach to sobriety has never made sense to me. From our family backgrounds and the communities we grew up in, to our physical attributes, the education we received or didn't, and the media we've been exposed to, we will all have developed a different relationship to alcohol—along with our individual reasons to believe in the bliss-inducing properties of booze.

As you begin experimenting with longer and longer periods of abstinence, try to develop an interest in the specific situations where FOMA kicks in for you. On the other side of the knee-jerk response to reach for a drink, this is valuable insight into all the reasons you use booze: aka Sober Curious gold.

For me, it was holiday visits with my family, weddings, bachelorette and birthday parties, anticipating a vacation, any kind of restaurant situation, attending a media networking event, especially if I knew I might have to speak to a "VIP" (not that uncommon in my last magazine job, as I was responsible for compiling the weekly "party page"—which basically meant going to the opening of every envelope in town and attempting to get a "funny" quotation from any celeb in attendance. This was the natural evolution of the "girl about town" status I craved in the early days of my career, I guess, but pretty much my idea of hell).

You will no doubt have spotted something of a theme: I feared I would miss having alcohol as my ally in pretty much any and all social interactions that occurred outside office hours or home alone with the Pisces. How would I start a conversation? Would I be dressed wrong? Exactly how awkward would the small talk be? Having gone and said all that about our unique attachments to booze, I have to add, given alcohol's ability to *switch off the part of our brain that monitors how we are perceived by others* (more on the specifics of this in the next chapter), that these fears are not at all uncommon.

It's the moments of euphoric recall that are actually more specific to each of us. For me, these can swoop in as suddenly as a tropical storm and feel almost like an out-of-body experience: when the idea of "a drink" becomes not only filled with pleasurable anticipation, but is drenched with nostalgia for *the best days of my life*—tinged with melancholy that those days may now be in my past. Passing a bar in a train station or airport at the start of a long trip; the "tools down" moment that marks the meeting of a deadline or completion of a project; the specific way the light hits the buildings on the first warm day of spring; the smell of a cigarette on a hot street; the moon being in the sign of Leo; a precise style and cadence of deep house drum beat.

So here we have two different kinds of triggers: one motivated by fear of social discomfort, the other by a craving for an experience of something like magic or transcendence. Both of which were equally likely to make me reach for a drink as I navigated my new hangover-free existence. What are yours? Thinking back to Marc Lewis's theory that all human behavior stems from our desire either to seek out pleasure or to avoid pain, it seems obvious that our specific FOMA triggers will be individual for each of us, even if they are rooted in the same basic needs.

It's also important to note that none of my triggers is linked to an especially traumatic life event—since it's the point at which alcohol becomes less social lubricant and more self-administered anesthetic that the hook of addiction can sink deeper into a person's brain. It could be considered another "privilege"—the same way I get to choose to be Sober Curious—that I never learned to use alcohol as a way to cope with a painful loss or a period of depression, for example. A privilege first and foremost since I have experienced relatively few traumatic life events—but also since my background gave me the resources and confidence to trust in my own ability to cope when things get tough.

In the HBO documentary *Risky Drinking*, which profiles, well, risky drinkers from all different walks of life, experts discuss the "shades of grey" that are now widely acknowledged to exist when it comes to addiction. At what point does grey become black and white? At what point does the amber warning light of addiction begin to flash red? "I like to say it's when alcohol becomes a friend, the thing that we turn to to relieve our stress, to numb our pain," says Deidra Roach, of the National Institute of Alcohol Abuse and Alcoholism. "Because when alcohol becomes a friend, it's firmly on the path to becoming a partner. And as a partner, it's poison." One way to read this is that the more severe the stress, the more traumatic the pain, the more likely it is that "normal" drinking will slide stealthily into becoming risky drinking.

This gets deeper into the roots of our experience of FOMA. Perhaps what so many of us really fear about missing alcohol is coming face-to-face with the shame, the discomfort, the absence of joy, and in some cases, the abject pain that make us drink in the first place. (Something we'll also investigate further in another chapter.)

For now, the key to working productively with euphoric recall (as with so many things in life) is *balance*. For example, for every effervescent memory of dancing on tables and bonding with new BFFs, I can also choose to bring to mind a morning after spent projectile vomiting each time I tried to take a sip of water. Contorting my body into agonized shapes in a tangle of sweaty sheets. A close call with grievous bodily harm brought on by my drinking. The creeping existential crisis of a three-day hangover.

I would be lying to myself if I didn't acknowledge *both* sides of the coin. The sunshine *and* the sorrow. And while not every experience of drinking led to wretchedness and regret, my honest truth is that overall, the negative consequences far outweighed the good times.

In AA they call it "remembering your last drunk"—one of many useful tips for living sober in a slim volume of the same name published in 1975 by the organization (which is not just for alcoholics, by the way, and which anyone can order from Amazon). "This is not a typographical error," the book explains. "The word is 'drunk,' not 'drink' . . . 'A drink' is a term which has awakened pleasurable echoes and anticipations in millions of people for centuries." As soon as euphoric recall creeps in, the advice is to "think the drink all the way through, down to our last miserable drunk and hangover." It's what they also call *playing it forward.* And in the case of conquering FOMA, it's one of the most effective tools in the box.

How Am I Supposed to Socialize Sober?

This has got to be the number one question at our Club SÖDA NYC events, and in fact the reason a lot of people attend in the first place is that they're seeking "sober" social buddies. Anyone who's felt like the lone nerd nursing a soda and lime while all around you people revel in the ecstasies of alcohol will be familiar with how that often goes: the awkward moment of the first refusal; the lame excuses ("I'm on antibiotics"); the subtle or not-so-subtle "encouragement" to "join in"; the feeling more and more like an outsider/everyone's mom as the volume gets dialed up, the conversation begins to veer all over the place, and eventually you take yourself quietly home.

One other thing I want to talk about here, though, is the importance of facing Sober Firsts, that is, your first time attending *X* kind of social event sober, head-on. And it's capitalized because Sober Firsts are something I talk about a LOT, as they are a big part of any Sober Curious journey.

The temptation, when FOMA kicks in, may well be to (1) cancel your social life until further notice and/or you've worked out how not to drink without feeling like a total loser, or (2) decide you're going to drink only in situations where it feels like it will be impossible not to. As I said, the big ones for me and many people I know are things like weddings, birthday parties, dance parties, any other kind of party, Christmas day (and other holidays where drinking is the norm), vacations, work bonding sessions, first (and second, third, fourth, etc.) dates.

The problem with option 2 is that if you drink "only" on any/all of such occasions, you'll likely wind up drinking most months. Maybe even most weeks, if you have a lot of friends, are actively dating, and have plenty to celebrate (which is the kind of life we're aiming for here, right?). And drinking most weeks, or even most months, is basically the opposite of the point, which is, if you recall: *that the only way to change your drinking habits . . . is to change your drinking habits.*

Since option 1 is also a fast track to feeling sorry for poor-old-Sober-Curious-me (not to mention that human connection, as I've said, is *by far the greatest predictor of overall happiness*), the only way forward is really (3) to RSVP "yes," go boldly forth, stripped of liquid courage and beer goggles, with no pregaming whatsoever, and see what happens.

You will likely feel some discomfort at first. After all, you're going against neural pathways that have been shaping your brain since the day you started drinking. Depending on numerous factors—the people involved, your general mood, whether there's any food—this discomfort may last right up to the point you call it a night. But in my experience, you're more likely to discover that once the initial wave of "WTF" has subsided, it's totally possible to have a very nice time with these smiley, happy people when you're sober. For a few hours at least.

Sitting in the WTF, watching it pass, and then choosing to focus on the positive parts of the experience is the physical part (and the mental workout) of your beginning to create *new* neural pathways. And this is really how you change your drinking habits. I call it "getting comfortable with being uncomfortable"—a life hack (as we'll discover later) that can benefit you in so, so many ways.

Sober Firsts aside, the bottom line when it comes to sober socializing is that if most of your social life has been happening in bar-like settings, getting Sober Curious means this is likely going to change radically. By all means, still hang out in bars, but these environments haven't exactly been designed with nondrinkers in mind. For starters, the conversation has got to be *really good* if there's no other activity besides drinking to engage in—and one myth my own Sober Curiosity has been quick to debunk is that conversations are better/funnier with booze. They're not. Often, it's just that our boredom threshold goes up.

This is also very helpful when it comes to determining *who* you want to hang out *with*. Lack of chemistry (with friends and lovers alike) becomes quickly evident without the forgiving veneer of a drink, and your friendship group may also shift to reflect this. Which is perfectly okay. We're taught that a real friend is a "friend for life" (whether this applies to us or them)—but in truth, the people we feel closest to and who inspire us most with their presence can change dramatically over the course of a year, let alone a lifetime.

There's no need to make a drama out of it, either. You'll likely find that your change in drinking habits leads to a natural reshuffling of your #Squad, as you gravitate toward people whose #LifeGoals are more aligned with yours. I repeat: *This does not make you a bad person.* It makes you an honest one. A person who is living with integrity and therefore cultivating cognitive resonance.

Much of the above, by the way, is a reflection of *my* experience of getting Sober Curious. You might love the buzz of being the only sober person in the club, knowing you can stay out all night and still feel Fresh-with-a-capital-F the next day. You may also find your same drunk friends as hilarious and lovable when you're sober as you did when you were drinking with them. I have experienced all these things too, and having a night out when you laugh and dance nonstop AND are able to get up and kill it at the gym in the morning feels like double-dosing on life.

A word of warning, however. Having gotten Sober Curious and discovered all the positive side effects, you may also experience what is referred to in recovery circles as "the evangelical phase." You'll know this has kicked in when you begin seeing it as your mission to show all your old drinking buddies the error of their ways. After all, you feel so great, why wouldn't you want to spread the love?! This includes positive side effects of not drinking, such as (and you can consider this a little motivation as you contemplate your potentially quieter and perhaps-slightly-awkward-at-first new social life):

Experiencing regular, deep, properly restorative sleep

Having improved digestion

Having clear skin

Feeling loads more confident

Being way more productive

Feeling less anxious

Experiencing more optimism, energy, and excitement about life in general

Being at massively lower risk of developing multiple chronic diseases

Wanting to help others more

Saving a bunch of money

Sleeping only with people you really want to sleep with

But a word of advice: DO NOT make it your mission to show all your old drinking buddies the error of their ways. This is the fast lane to judgment city. Your Sober Curious journey is yours and yours alone. How other people choose to waste their weekends, flush their youth down the toilet, and poison their bodies with toxic chemicals (lol) is none of your business.

Which raises another important point: It's very difficult not to come across as uptight and judgmental when you're not drinking and everybody else is. This is especially likely if you are "choosing" not to drink versus having a "problem" with booze or *actually* being on antibiotics (not that the latter ever stopped most of the drinkers I know from imbibing). Scenario 2 here, that is, having an alcohol problem, means the drinkers get to feel sorry for/judge *you*. But as Annie Grace writes in *This Naked Mind*, when your not drinking is a Sober Curious *choice*, "drinkers are unable to pity you, [and so] it's possible that they will begin to feel sorry for themselves. They may feel judged when you decline a drink. You are now holding yourself to a higher standard . . . This is great, but not everyone will see it that way."

And how many times has it felt like your not drinking is ruining everybody else's night?!

I think one reason I often remind people I'm not Sober *Sober*, that I still have a drink *on occasion*, is to mitigate the whole above judgment-fest, and to plant the subconscious seed that "I'm still one of you!" If it comes to it, it's also okay for me to accept and hold a drink, even take a sip with a toast, for example, and this not be termed a "relapse" or a hazard to my "recovery"; which makes it easier in certain situations for everybody else to feel better about my not drinking. Rather than my drinking "for them," I see this as an act of kindness and diplomacy.

On the other hand, it's also quite telling that I was always secretly in awe of the sober person at the party—the DJ friend who stuck to soda all night and stayed out as late as everyone else, the colleague who bypassed the wine at work events yet was still super well-liked. Looking back, it's obvious that part of me admired the fact that they didn't "need" a drink to fit in and feel at ease. I wanted for myself some of the quiet confidence they seemed to exude (and which is awaiting you just a little farther along your Sober Curious path, promise!). But admiration can easily turn to envy when whatever "they're having" seems like it might be unavailable to us also. Because of our garden-variety regular ol' social dependence on alcohol, for example.

Ultimately, our judging others often reflects something we're unhappy with in ourselves. Our own social awkwardness, perhaps, and the fact that we're willing to trade a few hours of what feels like effortless connection for the inevitable hangover. But when did being the life and soul of the party become the ultimate social currency? How did simply choosing not to drink come to get a person labeled a "killjoy"? And why do we associate "high times" with how many likes our party pictures get on Facebook?

Alcohol and the "Cult of Personality"

In her book *Quiet: The Power of Introverts in a World That Can't Stop Talking* (2012), Susan Cain writes, "We're told that to be great is to be bold, to be happy is to be sociable." She explains how the rise of what she calls "the Extrovert Ideal" has led to more introverted people thinking "*there is something wrong with me*" if they're shy, don't particularly like parties, and have trouble making witty repartee. Yes, I'm talking about me here. (Although since getting Sober Curious, I've also remembered that it's easy for me to be good at all these things, sober, when hanging out with people I really click with.)

I was really surprised, however, to find no specific mention in Cain's book about the role of this Extrovert Ideal in relation to addiction, and alcohol dependence in particular—apart from one scene where her husband gives her a water bottle filled with Bailey's Irish Cream to swig from before a big public speaking gig. Considering how many (as in *most*) of us use alcohol to feel more comfortable in social situations, the link between booze and our ability to attain Cain's Extrovert Ideal seems like an obvious one.

The biggest problem booze ever solved for me was that it helped me "come out of my shell" and act way more extroverted than I naturally am. Don't get me wrong—I love people, and connection is what makes me, as with every human, happiest. I just generally prefer people one-on-one and connecting with those I trust enough to be my whole self with—people who are actually few and far between.

The Extrovert Ideal meanwhile (not to mention social media) teaches us that having LOTS of friends, being liked by EVERYBODY, and juggling a packed social calendar that reflects this are the markers of a happy and successful social life. A successful life in general, even. And getting Sober Curious also means we get to question this.

Does everybody drink to make themselves feel more extroverted? Does having more friends make for better friends? What if many of us could benefit from more alone time than modern society says is healthy or cool?

Cain quotes what cultural historian Warren Susman called a shift from a "culture of character" to a "culture of personality" in twentieth-century America: "In the Culture of Character, the ideal self was serious, disciplined, and honorable. What mattered was not so much the impression one made in public as how one behaved in private." With the advent of the culture of personality, meanwhile, "Americans started to focus on how others perceived them . . . [as] Susman famously wrote: 'Every American was to become a performing self.'"

Based on what we've learned about the relationship between booze/addiction and feelings of shame (*there is something wrong with me*), as well as the specific role alcohol plays in numbing how we think we are being perceived by others, it's easy to see a correlation here. The more pressure to conform to the Extrovert Ideal, the more tempting it becomes to take a drink to ease our "performance anxiety."

Social media plays a very particular role in this. On one hand, you could say apps like Instagram and Facebook exacerbate the Extrovert Ideal. Cain's book came out *before* 2014 was declared "the year of the selfie," arguably a tipping point in terms of social media being used less as a way to keep up with friends and family and more as a tool to create and promote individual personal brands. Now that we also have Instagram Stories and Facebook

Live, these platforms give each of us the opportunity to be—some (those feeling the pressure of the Extrovert Ideal most keenly, for example) might say *insist* we be—the stars of our very own reality shows.

But on the other hand, these apps also make it possible for socially awkward introverts to seek out and pursue connections with like-minded individuals without having to navigate the noise of bar culture and cocktail parties. When they are used consciously, that is—since social media is also often cited as being at the heart of the "disconnection epidemic" that Brené Brown writes about in her books. When we mistake follows, likes, and comments from random strangers for the real thing, forsaking real-life connections for perceived digital kinship, this can definitely be the case. But with careful curation, it is possible to create thriving social connections online that require no awkward small talk, or the drinks to massage it, whatsoever—something I often mention when asked why I think that more and more of us are getting Sober Curious.

Which is not to say that only introverts benefit from the social lubrication of alcohol. For more extroverted souls, the ones who love being seen and who recharge their batteries by hanging out in groups, booze is just fuel for the fire. For somebody who loves the buzz of being around lots of people, why would they ever want the party to end? If anything, FOMA can be worse for extroverts, since getting drunk will likely have become a by-product of something they are naturally drawn to—people and parties—as opposed to introverts using booze to make it easier to be social.

Being more likely to RSVP "yes" without any internal wrangling, extroverts will likely experience more awkward Sober Firsts in quick succession, more being the odd-nondrinker-out. They may have to put in more legwork finding situations in which to socialize sober, while perhaps feeling even more disillusioned with the dominant drinking culture.

Ultimately, this just means all the more opportunities for mental, emotional, and spiritual growth!—even if, at first, the question surfacing time and again is one of the most daunting of all: *Exactly how boring will my life be without any alcohol in it?*

Life Is the Opposite of Boring Without Booze

The clue here is in the lingo, the perception that while drunk people are out "partying," sober people are sitting at home in food-stained slouchies being "dry." The implication being that living Sober Curious will be dull as dishwater. Of course, as discussed, this assumes that parties are your idea of fun—definitely not the case for everyone. But as I contemplated life without alcohol, FOMA also surfaced when I considered just how *monotonous* life was going to become. How very *one-note.* After all, as one of my old magazine editors very succinctly put it: "*Without Friday night drinks, how do you know the weekend has begun?*"

My favorite sobriety memoir is *A Nice Girl Like Me: A Story of the Seventies* (1984) by another British journalist named Rosie Boycott, who was also the cofounder of the 1970s feminist magazine *Spare Rib.* In her memoir, she asks a doctor in her rehab facility: "How on earth is life ever going to be exciting without booze? Booze is associated with excitement, with good times, with bad times, and somehow the thought of life lived on an even keel is very depressing." Which sums up another element of my old FOMA-infused thinking (and begs another question): *Surely the lows were worth it for the highs? Without the skull-crushing hangovers and the existential crises, would the dancing on the tables ever happen?* (I'm speaking literally *and* metaphorically here).

Imagine my surprise, then, when after a few weeks of not drinking I often found myself doubled over in hysterical laughter for no "real" reason—other than the surge of sheer *joy* I got from seeing a ridiculously cute new picture of my nephew on Instagram. Or skipping down the street because of how happy it made me to get a cute text from the Pisces. Feelings of joy I realized I'd been suppressing, *along with* the daily anxiety, restlessness, and discomfort I used alcohol to allay. Anxiety, restlessness, and discomfort that Annie Grace suggests are also *symptoms* of a dependence on alcohol.

Booze is essentially an anesthetic—and another favorite Sober Curious quotation, which we use in our Club SÖDA NYC literature, is this one from Brené Brown: *"Numbing vulnerability also dulls our experience of love, joy, belonging, creativity, and empathy. We can't selectively numb emotion. Numb the dark and you numb the light."*

Unexpectedly, as my Sober Curiosity progressed, a more sinister pattern also emerged. The more naturally joyous, exuberant, and buzzed I felt on life, the "higher" I got on my own supply, the stronger my cravings for a drink became. Partly because I associated feeling this way with being drunk, I guess. Because my euphoric recall was triggered. But also, I came to realize, because these feelings felt as volatile and uncomfortable to me as the miserable ones. I wasn't used to feeling excited about simply being alive, feeling optimistic and confident for a positive outcome—whatever. In fact, for multiple reasons relating to our specific life circumstances, I would go as far as to say that *most of us are not used to feeling this way.*

If you're a regular drinker, the alcohol itself (or rather, the lingering toxic aftereffects) could be the reason. And if you're used to drinking to feel good, alcohol highs will have become way more familiar and predictable, and therefore easy to manage, than the ecstatic full-body chills you may find yourself experiencing on a random Wednesday morning—just because you're a happy, healthy human, and also, perhaps, as a result of your not drinking. Out of context, these "highs" can feel pretty precarious. A little manic, even. They might just, heaven forbid, compel a person to kick anything to the curb (shitty job, unhealthy lifestyle, controlling partner, etc.) that is getting in the way of his or her feeling this way more often. If you hate your job, for example, the "escape" of Friday night drinks means you have to endure only five days of misery at a time. You'll quit that job and do something more interesting with your life *way* faster if there's no get-out-of-jail-free card waiting at the end of the week.

As for what to do any time you feel blindsided by these overwhelming new *positive* feelings? Breathe. *Sit in the WTF, watch it pass, and focus on the positive parts of the experience.*

Contemplating the "middle of the road ordinariness" of a life without alcohol, Rosie Boycott writes in *A Nice Girl Like Me*:

"Sure, booze was providing my excitement, because it was inducing chaos. I'd be swinging into a bender, out of it, through a period of furtive but controlled drinking and then back to chaos again by getting almighty drunk. I'd forgotten, in the interim, that whatever feelings of happiness or sadness I had were just as powerful when left alone. More powerful, *because they were real.*"

Put that in your FOMA pipe and smoke it. Boring? Life without booze is anything but.

Sober Socializing Checklist

- Switch up late-night dinners for daytime brunch dates. Added bonus: The gossip gets even better when you can actually hear it.
- Choose social situations that also include an activity: yoga, bowling, comedy shows, talks, and crafting are all good options. Yes, I said CRAFTING. One Sober Curious friend told me that making something with her hands satisfies the exact same part of her brain as a drink!
- Just get past the first round. How long it used to take my brain to trust that the ground wasn't going to open up and swallow me because I wasn't drinking. Also, how long it took most of the drinkers to get buzzed and forget to bug me about it.
- Don't be embarrassed to ask for an alcohol-free beer, a fancy mocktail, or a tonic with bitters—they're way more fun than soda and lime (check out "What to Drink When You're Not Drinking," page 190). (NB: "Non-alcs," as I like to call booze-free beers, are often frowned upon in recovery circles—even the 0.5 percent alcohol they contain can be considered a relapse. Ditto kombucha. But for many Sober Curious souls, they're game changers!)

- Order dessert. And savor every bite. Only after I got Sober Curious did I realize how many of my alcohol cravings were also sugar cravings—and now is not the time to pressure yourself to quit the sweet stuff too (more in chapter 5).
- Remember that your choice is a privilege . . . and also that it's not your job to judge others. If you find yourself going there, stop, observe, and ask what their behavior is triggering in you.
- See every Sober First as a chance to grow new grey matter (i.e., literally get smarter! Marc Lewis's research has shown that this is what happens when we go "against the grain" and forge new neural pathways).
- Revel in the feeling of waking up hangover-free. Taking time to appreciate (definition: *increase in value*) this warm, radiant, yes, okay *smug*, full-body glow will help train your brain to want more of it. Years into my Sober Curious journey, I would find myself feeling grateful for not having a throbbing, groggy head whenever I woke up in the middle of the night to pee!
- Above all, gentle with yourself. AA preaches "one day at a time," and when it comes to conquering FOMA, the same goes for taking it one dinner, one dance party, and one date at a time, too.

SOBER CURIOUS LOVE, SEX, AND RELATIONSHIPS

If you are a regular drinker, somebody whose social life revolves largely around booze, and for whom dinner dates, parties, and family holidays are awash with the stuff, then chances are that alcohol also plays a pivotal role in your closest and most valued relationships.

It's how you bond with your ride or dies. It's the secret sauce that spices up your long-term comfy-but-kinda-too-cozy-for-its-own-good partnership; a sickly-sweet glaze to smooth over family differences at Thanksgiving; the lube that keeps your sexy singleton status well-oiled.

Sometimes, you may even have mistaken the companionship of a robust glass of red for a lover in and of itself—when dining out alone, perhaps, and a book feels less like company and more like a conspicuous signal that *nobody loves me*; when you're curled into the sofa, solo on a Saturday night; or when you're all out of Tinder swipes and are craving (from the very depths of your soul) a cuddle.

If one of the most common barriers to conquering FOMA is fear of sober socializing, the thought of removing booze from our intimate relationships—not to mention the behaviors we engage with in the hope of meeting, and potentially leaping into bed with, the loves of our lives—can be daunting as fuck. *What will we have to talk about? Will I ever feel relaxed enough to want sex? Confident enough to ask for it? How well do I even know these people? Will we still love each other sober?*

As Ernest Hemingway famously put it, "I drink to make other people more interesting." To which I can personally add, to make other people seem more attractive, more lovable, more caring, and more like they really "get" me. As I touched on in the last chapter, getting Sober Curious often leads to a reevaluation of the people you feel most comfortable being vulnerable with. Including being physically naked with. And if I made it sound as if this would happen naturally—like the changing of the seasons?—well, at times it may feel more like natural *childbirth*. Which is to say laborious, and not without its moments of contraction.

Because what if the person or people you do most of your drinking with are the ones who are always there for you? The friends and lovers who make you feel truly lit? Or the partners you share your home, have made your babies, and have mapped out your retirement plan with? What if breaking up with booze could also mean breaking up with them?

First of all, this is a worst-case scenario. To imply that *all* your most loving relationships have been built solely on a shared appreciation for bottomless mimosas would be ridiculous. True love, connection, companionship, and chemistry will endure, regardless of your brunching habits. But when you first get Sober Curious, where these qualities have also been amplified by ample alcohol consumption, there may be a period of adjustment. Okay, a few periods. Days, weeks, and months, even, requiring plenty of patience, truckloads of compassion from all parties, and, ideally, the support of some fellow Sober Curious allies.

And can we zoom out here for a second? What does it say about our attachments to alcohol when the thought of removing booze from our lives often leads immediately to our questioning who our real friends are? Whether we're going to have to choose between *it* and them. Could it be you've got a closer relationship to alcohol than to some of the people in your life?

It's a sobering thought. And one I have most definitely been confronted with.

As you've already read in the intro, the beginning of my love affair with booze coincided with my diving head and heart first into an actual love affair with my future husband, the Pisces. My new life with this intoxicating new person revolved around pubs, clubs, literal tequila sunrises, and googly-eyed dinners that *always* ended with an Irish coffee.

It being early noughties London, Kate Moss falling drunk out of members' clubs was my ultimate "girl about town" role model. Pisces over here was a *party promoter*, FFS. Did I want to admit to my new lover (or myself, for that matter) that I was a secret geek? Would often have preferred to go to bed early with a good book? I did not. For a long time, it also felt transgressive and wild and ecstatic to explore a part of myself that I'd glimpsed only in movies and magazines. To be Sweet Valley High's Jessica, for once, instead of studious Elizabeth. Alcohol also helped me feel like a far more "exciting" romantic prospect than I considered myself, deep down, to be.

The cracks began to show when we hit the classic seven-year itch. In fact, years before I got fully Sober Curious, reevaluating the role of alcohol in my marriage was when the very first questions started.

Sitting in our local pub one standard Friday after work, the large glass of Cab Sav in front of me appeared to have uncorked a steady drip of tears. I couldn't pretend anymore. Somebody had removed the sexy devil mask from our marriage to reveal plain old Scooby Doo underneath, and I was *devastated*. To my surprise, I found myself nodding at my drink and asking, "What if *this* is all there *is*?"

So melodramatic! In my defense, I'd spent the majority of my adult life thus far avoiding dealing with my more complex feelings in a mature and intelligent fashion. Fast-forward to now, and twenty years into our union I've learned that all relationships ebb and flow, require constant caretaking and renewals of commitment. Not to mention that the emotional clarity of sobriety is exceptionally helpful with this.

As for the Pisces? What he did not do (and herein lies the moral of this story) was react like I was trying to start a fight. He did not take my mini meltdown personally, and he chose to hear me out. He went along when over the coming weeks, months, and years I suggested a Sober Curious date night (awkward and clumsy at first, as neither of us had been out after dark without liquid assistance since we were teenagers) and refused to feel judged or get defensive when my drinking choices shone a light on his.

He's been a superstar, basically. And with no bullying, threats of divorce, or other forms of coercion from me, these days he's Sober Curious too.

Which is jumping ahead of ourselves a little.

Because before cozy Sober Curious domesticity sets in, there comes the question of Sober Curious dating.

I'm Supposed to Date Sober, Too?

If the first thing people want to know at our Club SÖDA NYC events is how to socialize sober, this is often VERY closely followed by the above line of inquiry—perhaps because we're based out of New York, widely regarded as the complex-inducing dating capital of the world. But also because sitting across from a stranger sober (meaning emotionally *naked*)—a stranger you may also be angling to have sex with, escape as fast as possible from, or perhaps even share the rest of your life with—has got to be up there with public speaking when it comes to the social situations we're most terrified of.

Here are some of the comments I got when I canvassed my crew about their experiences of Sober Curious dating (necessary, because I haven't technically been on a date with a new person since 1998, and the overall logistics of looking for love have shifted *radically* since then):

> **"I don't want to typecast myself as a 'sober' person on a first date."**

> **"Most girls look at me like I have two heads when I tell them that I don't drink, and nine times out of ten they start prying into why."**

> **"It can be really confronting and daunting. 'Do they think I'm pretty? Does he smell good?'—a glass of wine quiets all the stories."**

And also:

> **"He thought it was attractive and admired me for the courage to make a lifestyle change when that lifestyle wasn't working for me."**

Apps like Tinder and Bumble have given rise to an era of casual, often booze-fueled hookups, but the basic human needs and insecurities at play remain the same. Whether canvassing for a potential life partner or a plain old fuck buddy, we want to be perceived as smart, confident, and most definitely not as if we might be hiding an extra head, might in some way be shamefully *other*.

We got into all this at a SÖDA event titled "Sex, Lies & Alcohol"—which at the time drew our biggest turnout to date, despite it being the middle of August when the myth is that most New Yorkers are off swigging rosé in the Hamptons. Turns out they're not. They're still running around the city, neurotically swiping right (or not, and instead pondering how the hell you used to meet anyone before there were apps for that).

My SÖDA cofounder, the spiritual teacher Biet Simkin, had this to say about drunk dating: "If you're in it for anything serious, it's the same as going on a job interview. And would you get wasted before that?" The implication being that not only will you not be presenting yourself in the best possible light, you'll also walk away with little idea as to whether you negotiated for what you're worth. Let alone be in any fit state to size up the real potential of your date.

One Sober Curious friend got set up with a guy by some friends when she began sober dating and was glad not to be drinking because "he was making all these stupid comments and made me feel really small. The messed-up thing is, if I'd been drinking I probably would have gone home with him, and then spent the next few weeks stuck in an awkward texting thing." In the event, she cut him off that night—saving a ton of time, energy, and heartache.

The "Lies" part of our discussion focused on the specific ways in which alcohol helps us disguise the character traits we may find less attractive about ourselves, and therefore assume will make us less attractive to others. Also, the lies we buy into when we're under the influence. For example, the untruth that alcohol makes us "edgy" or "cool" (when actually we're conforming to the societal norm), that it makes us cuter (I take it no one here has ever seen a photograph of yourself drunk, with bloodshot eyes and food on your shirt), that it relaxes us (when really we just pressed "pause" on the stress), that it helps the conversation flow (when, as already established, it just raises our boredom/BS threshold).

And the big one: that alcohol is what gives us the confidence to be ourselves—when all it really does is suppress the right temporoparietal junction (the *official* name for that part of the brain that monitors what other people think about us) while exacerbating risk-taking behaviors.

Oh yeah, about how alcohol messes with our judgment? When I took the written test to get my New York drivers' permit, at least half the questions were related to the dangers of drunk driving. How operating a car (and any other heavy machinery for that matter, like, say, *your body*) while under the influence makes you a danger to yourself and others. Imagine if we got similar messaging about drunk dating, the "car-crash" situations that could be avoided!

Which it's all helpful to be aware of, but which doesn't make the actual prospect of baring your soul/body to a stranger whom you want to encourage to like you any less unnerving. I consulted with Alexandra Roxo on this one, a thought leader on love and connection in the digital age. "It's terrifying to sit across from another human and to ask yourself to open, because what we're faced with is, 'How do I know I can trust you?'" she explains. "Alcohol helps boost dopamine production, which is a heart-opener. And connection cannot take place if our hearts are closed for business."

As for how to facilitate this opening, or trust, without a little *liquid courage*? Entry level, Roxo says, is to "breathe deeply and slowly into your belly, subtly synching with your partner's breathing. Remember to make eye contact every so often. Allow yourself to laugh and be a little silly. Then it's about asking questions that promote vulnerability without being intrusive, like what song a person has had in their head all week and why they love it."

Another tactic Roxo points to is to "take a risk together, by trying an activity neither of you has done before. This is another way to shift the chemical balance in your bodies to create an opening and a mutual vulnerability." Which could be as extreme as a bungee jump, or as simple as . . . both of you not drinking. Considering this will be a Sober First for a lot of people, stating up front that you'd rather not drink and instead take a walk, meet at a museum, or check out a fancy ice-cream parlor can feel "risky" in and of itself, given that it's so outside of the norm.

In the old days, this stuff might have been called *getting to know someone.* And if dating Sober Curious also means things will likely go a lot slower, then it's important to remember that "taking your time to really discover another person is what creates true intimacy," Roxo says.

And speaking of intimacy . . .

We All Know Sober Sex Is the Best Sex

I wish I could drink like a lady.
"Two or three" at the most.
But two, and I'm under the table—
And three, I'm under the host.

—ATTRIBUTED TO DOROTHY PARKER

Whether you want "it" to happen that night and be a one-time thing or are actively interviewing for a long-term union complete with Netflix, "chilling," and carefully mapped-out baby-making, going on dates is often in service of facilitating sex. Depending on your personal needs, the fact that a technology-driven hookup culture can make it seem like that's *all* it's for is either a sad indictment on modern life or a shortcut to sexual nirvana.

But one thing is for certain. "When a person's mind is disengaged and they're more relaxed," Roxo says, "the illusion is that the booze is making them feel sexy. The booze is not. The lack of thoughts and the lack of insecurities is making them feel sexy. Then the alcohol is also creating a lack of sensation, and a lack of emotional connection." Or as my Sober Curious friend Kate puts it, "It's ironic how everyone drinks to have sex . . . and then the sex is shit."

Hands up, ladies, if you have a harder time climaxing when you're drunk. And you too, dudes, if you've ever experienced "brewer's droop." As well as alcohol being the equivalent of rubbing numbing cream on your genitals, a 2007 study published by the National Center for Biotechnology Information showed that drinking is the leading predictor of a person developing sexual dysfunction. When it comes to the question of sober versus sozzled sex, I think we all know the real answer. Not to mention that drunk sex is also less likely to be safe sex, with a study by Johns Hopkins University showing that women who binge drink (five drinks or more at one time) are five times more likely to contract gonorrhea than those who abstain.

It's also telling that the cozy three-way the Pisces and I found ourselves in with alcohol never extended to the bedroom. In twenty years of marriage, I can probably count on two hands the number of instances that we've had drunk sex—most of them in the first couple of years of our relationship. Beyond that, our deepening emotional intimacy has bred deeper sexual intimacy, which alcohol only hinders.

But if sex with a new person, or with a partner you're used to getting buzzed with before you do it, is also a Sober First for you, then it can be one of the most intimidating of all. (Which, good news, means it also provides the most opportunity for mental, emotional, and spiritual growth.) The personality flaws we're not so fond of? It's easier to reveal those incrementally as you navigate the "getting to know you" phase. The physical ones, including any performance anxiety around taking a long time to come or failure to get it up (for example), less so.

When the majority of your intimate relations have been happening while under the influence, the other good news, however, is that extending your Sober Curiosity to your sex life means things in this department are about to improve dramatically. Even if at first it might mean having less sex.

But since when did *more* equal *better*?

When I was around nineteen years old, I asked a friend I looked up to as an older sister how often she and her long-term boyfriend had sex. She told me about once a month—and that she always had an orgasm. Well, my then-current "beau" (yep, that old Capricorn) insisted we do it three times a week, minimum, and there were *never* any fireworks for me. The messed-up thing is that all the messaging I'd internalized by that stage in my life—from advertising, movies, women's magazines, and the porn industry—told me that *I* was the one with the better sex life. If getting Sober Curious also means facing some sober realities, rewriting the script that more sex equals better sex may well be one of them.

But back to the actual moment you're about to do the deed and there is no boozy veil behind which to hide your awkward, lumpy, imperfectly perfect human self. The first thing, Roxo says, "is to acknowledge the nerves and anxiety. To know that it's okay to say to your partner, 'I'm feeling really nervous.'" Which is about as vulnerable as it gets. *Especially* with an existing partner, somebody who supposedly already knows you intimately. If sex usually "just happens" as a result of your falling into bed with each other after a few drinks, it can feel like removing that sexy devil mask to reveal *plain ol' me* underneath. This could be one of those renewal-of-commitment moments required in the caretaking of a long-term relationship—and the potential, when approached with an open heart and mind, is for it to be an intimacy accelerator. With a new person, meanwhile, the response you receive to a statement like this can also be an excellent indicator as to whether this is really somebody you want to share your physical and emotional insides with.

After all, aren't the shyness and inhibitions we're taught it's "liberating" to wash away with alcohol, and that Susan Cain says we've developed a cultural aversion to, actually nature's way of helping us practice *discernment*—not to mention, considering the role of both internet porn and dating apps in a culture where being sexually available has become part of the Extrovert Ideal, another element of our humanness that's being steadily eroded by advances in technology? The message from both the online porn industry and the app-fueled hookup culture is that to be sexually confident and highly sexually active are part of what equals success as an adult human.

Given the complexity of our sexual identities and needs, opinions on this will be different for everyone. *As individual as our fingerprints*. Some people need more sex than others and will be more up front in asking for it. Others need an emotional connection to be established before sex. But if the conversation is happening while one or both of you is out of it—meaning not fully present—lines can become blurred.

Which unleashes a whole other can of tequila worms.

The Role of Booze in a Hookup Culture

That supposedly-Dorothy-Parker rhyme above? It might have seemed witty and risqué back in 1959 (when it appeared in a magazine at the University of Virginia), but in the post-#MeToo era, we can't have an honest conversation about sex, lies, and alcohol without acknowledging the role of booze in "rape culture"—a term first used by feminists in the 1970s in an effort to shift consciousness around the inherent misogyny and sexism of "traditional" gender roles and how they normalize and trivialize sexual assault—namely, that "men" are sexually confident and dominant and "women" are sexually submissive. Our conditioning has created a culture where many instances of sexual violence still get written off as "men being men," where women have been taught to "lie back and think of England," and where victims of sexual assault are repeatedly accused of having "asked for it."

The role of alcohol in this dynamic? I believe booze not only exacerbates but also enables this toxic messaging. The fact that it bolsters sexual bravado while impairing judgment and amplifying risk-taking behaviors plays a major role in how we see ourselves as sexual beings—how the hookup culture fueled by porn and dating apps encourages us to "perform" these roles and thereby increases the potential for harm, to ourselves and others.

As well as increases the emotional dis-ease of feeling like in order to be a "desirable" sexual being, we must conform to certain narrowly drawn societal norms.

Which is not to sound alarmist, to suggest that without alcohol there would be no sexual assaults, or to imply that anyone is ever at fault for being assaulted. Rape happens because of rapists. Nonetheless, alcohol is recognized as being the number one "date rape" drug. A 2007 study by the National Institute of Justice reported that 89 percent of "incapacitated" sexual assault victims said they had been drinking alcohol or were drunk before they were assaulted.

Bizarrely, I also think the role of booze in third-wave feminism, with women learning (or rather, being encouraged) to drink not "like ladies," but "like men," is partly responsible for this unhealthy state of affairs.

In the United Kingdom, the late 1990s and early 2000s was the era of the "ladettes"—young women who were a by-product of 1970s and 1980s feminist thinking regarding the equality of women. According to a 2017 article in *Vice*, a ladette was "mouthy, up for a laugh, took her clothes off and could out-do any male companion in the drinking stakes . . . [The ladette culture] took the old, stuffy gender model—where the man went out boozing and the woman patiently waited up—and flipped it."

On the other side of the Atlantic, *Sex and the City* presented a far more well-groomed posse of cocktail-loving sexual libertines. Claiming our place alongside our male colleagues as we spilled into the SoHo streets after work, when my friends and I weren't downing "macho" pints of beer, our drinks of choice were Samantha and Carrie's pleasingly pink Cosmopolitans and Sea Breezes. (Good thing the cranberry juice in both helps with UTIs.) The underlying implication in both examples being that booze has come to play an integral role when it comes to a woman being seen as attractive and modern (meaning sexually liberated).

I was working on that trendy magazine at the time (the one run by gangsters who paid us in cocaine), and the (female) editor was also doing consulting work for some of the major drinks brands. The alcohol industry had identified a market opportunity—namely, that women still did not drink as much as men—and she'd been hired to help the industry better understand how to sell its products to the female of the species.

Annie Grace, a former marketing executive herself, included in *This Naked Mind* a very important fact that relates to selling alcohol specifically to women: "Marketers actually create need by speaking to your vulnerabilities . . . [and promising] fulfillment, completion, satisfaction, and self-actualization."

Part of the message of third-wave feminism was not only working and earning money "like men," but drinking and having sex like them too. In the most basic and binary of terms, this meant drinking hard and fast and having sex confidently, devoid of emotion, and as often as possible—a dynamic that has been routinely exploited by the alcohol advertising industry ever since. The times when this made you feel kind of vulnerable? Like it might land you in a situation where some kind of *violation* was more likely to occur? Well, here's some liquid courage in the form of a nice pink sugary cocktail (or low-calorie, carb-and-gluten-free shot of "skinny vodka," if that speaks louder to your body-image issues).

I'm aware that I'm speaking in huge generalizations here. Obviously, not all alcohol use leads to incidents of sexual violation, and women being perceived as sexual *beings* as opposed to merely sexual *objects* can only be a good thing. There's also no shame in using alcohol as an elixir to help you pursue sexual encounters that may feel transgressive (to you personally, or in the eyes of society).

Having explored his sexual identity in the gay male cruising scene, for example, trans diversity and inclusion coach Aaron Rose (also now Sober Sober) told me how "people used alcohol and other substances to disinhibit and actually more fully embody sexual and intimate desires that fell outside traditional norms." Even if, ultimately, "alcohol blurs people's capacity to fully consent . . . it can create very dangerous situations. The sexual experiences I had under the influence were not the present, connected, and truly witnessing intimacy that I now experience in sobriety."

Because booze, people.

The way it helps erase our individual quirks (or *otherness*) as we seek to "fit in." How it helps us enact certain roles that fit the Extrovert Ideal. The exaggerated swagger, and self-centeredness it enables. The numbing of our more vulnerable, complicated feelings. For every step toward true sexual equality, and intimacy, between human beings, the innate qualities of our social elixir of choice can't help but seem a little *regressive*, particularly as it relates to toxic masculinity.

In the 2017 Netflix documentary *Liberated: The New Sexual Revolution*, which follows college students on spring break, alcohol is the ever-present fuel for attitudes about sex in which "'making love' doesn't exist, it's not about emotions anymore"; where "ambivalence is the best attitude to have about sex"; and, more sinister, within "a culture in which sexual intrusion has become normalized." In the words of one guy, "All girls are panty-droppers—just give them a Percocet and a beer."

This comment pretty much encapsulates all that is problematic with the notion of alcohol and other drugs as a path to sexual liberation. Not that I'm here to judge anybody seeking a sense of personal freedom, relaxation, or pleasure. As we know, it's in our biology to pursue experiences that facilitate these things, and if anything, it evokes feelings of sadness that our lives are so starved of opportunities to let loose and feel free, let alone feel physical pleasure, that we're willing to risk personal injury to attain them.

That Percocet comment also reminds me of one friend's decision to get Sober Curious—how she came to this path having gotten into the habit of calling her dealer every time she got drunk... and having sex with him. At first it felt edgy and exciting. But after a while she began to wonder who was in control—whether the modern, sexually "liberated" woman in this scenario was really even her.

Getting Sober Curious Is About Learning to Love You

I will be true to myself and claim an authentic identity, not one scripted for me by the culture.

—MANIFESTO OF *LIBERATED: THE NEW SEXUAL REVOLUTION*

On the website for the 2017 documentary *Liberated: The New Sexual Revolution*, the filmmakers have included a belief system for how to #LiveLiberated from what they see as the shackles of the modern yet regressive attitudes about sex displayed in the film. Which, when you begin to look around, are still way more ingrained in mainstream culture than we'd like to believe. Especially in porn culture, absolutely in our attitudes about dating, and most definitely when it comes to the gendered roles around sexuality that we're encouraged to perform when we drink.

Considering what is actually the complex nature of our intimate relationships, romantic and platonic, and their role, ideally, as a place for us to both express and be accepted as our true selves, the above line is my favorite from the *Liberated* manifesto. And as it relates to our Sober Curiosity, "[alcohol] absolutely amps up existing cultural programming, as does anything that puts space between us and *genuine presence*," says Aaron Rose.

Alcohol helped me act more like the extroverted party girl I thought my partner wanted (let alone who would be accepted by my wider social circle). What "script" has it been writing for you? What role have you been playing in any relationship that revolves around booze? And exactly how much *effort* will it take to become genuinely present for the people you love, and who you would like to love and appreciate the "real" you?

Because you could say that alcohol kind of makes us *lazy lovers*—whether it helps us slip into well-worn gendered roles versus investigating who we are and what we want and need as sexual beings, has become a shortcut to intimate connection, or is a way for us to paper over the cracks (versus getting in there with the grouting tool) in our existing long-term relationships.

When I consulted sober self-empowerment coach David Wagner, who specializes in helping men connect more deeply with their emotions, he confirmed that "drug use, especially socially acceptable drugs like alcohol or cannabis, covers up a great deal of 'relational mismatch.' Rough edges can be smoothed over with a couple of beers; every conversation becomes 'deep' when you're stoned." In his work, he has also observed: "When sobriety enters a relationship and that artificial lubrication is gone, rough spots become more apparent. It's harder to hide. Many people find they are not as compatible or attracted or tolerant."

But on the flip side, "if the sobriety is approached in a conscious way," Wagner maintains, "a whole new level of intimacy can be opened. Hearts are uncovered. Revelations and 'deep insights' are legitimate and longer lasting." Yes, friends, just when this whole "sober relationships" thing was beginning to look like it might be all uphill, I bring you hope. Which doesn't mean it won't also take plenty of work—not least on your relationship with *number one* (yes, YOU).

How to begin? First of all, "Be gentle with yourself," Aaron Rose says. "When you remove alcohol from your system, it takes time to recalibrate. Sobriety has presented me with the challenge and opportunity to no longer seek instant external fulfillment of my desires . . . [It] has reconnected me with the knowledge that I already have everything I need to be safe and satisfied within myself." Meaning this is about to become a journey of investigating what, and who, truly makes you feel the most yourself. After all, aren't the best relationships (and the best sex) with the people who embrace and celebrate our unique, individual needs? People who actually find that extra head of ours kinda cute?

When I turned to my Club SÖDA NYC people again for some insights into navigating their new, emotionally naked love lives (whether single or attached), this is some of what they told me:

> **"When I go on sober stints, I don't go on a lot of dates. I see it as a time to work on myself."**
>
> **"If you love hiking, join a meetup group where you can meet a whole bunch of people who love it too. But know that first and foremost you're there for you."**
>
> **"Activism is a really good way of socializing. Every time I've been to a demonstration or a rally, there's a vibe. Like I feel like I could meet someone at a march."**
>
> **"Surround yourself with sober friends who share your interests. Check in with them before and after dates. Bring your intuition and your Higher Power along for the ride."**

The overriding message: The best way to meet someone, to reintroduce your partner to the Sober Curious you, or to work on your own *self-love* practice is to focus all your time and energy on the things you love to do—and in doing so, allow the people to do them with to find you.

Oh, and of the Sober Curious daters I quoted at the beginning of this chapter, who do you think wound up in a long-term relationship? The same woman whose date was most respectful of her decision not to drink also told me: "I 100 percent believe I am with a good man because I was sober and able to hear the intuition that led me to him . . . Compared to previous relationships it feels deeper, more authentic, and genuine."

Deep. Authentic. Genuine. Isn't this how we want ALL our

relationships to feel? Even if it means embracing the vulnerability of baring our souls, along with our bodies, and cultivating the patience to allow things to blossom and develop over time. After all, says Aaron Rose, "You've already done a huge brave thing—choosing to be fully present for your life. Celebrate yourself for that, and the rest will come."

Sober Curious Conversation Starters for Activating Intimacy

Here are some ideas to have in your back pocket for sober dates, postcoital moments, and cuddle sessions on the sofa. (What *not* to talk about? Anything you can learn about the other person from his or her Facebook profile!)

"What was the last thing that made you laugh out loud?"

"What would you like to be famous for? Why?"

"What was the most interesting thing you discovered this week?"

"Who do you get most nervous about calling on the phone?"

"What's the luckiest thing that ever happened to you?"

"What are some of your most vivid memories?"

"Who are you most grateful for in your life?"

"What do you value most about your upbringing?"

"What do you like the most about your physical appearance?"

"If you could discover one thing about your future, what would it be?"

SPIRITS AND SPIRITUALITY

Would I have gotten Sober Curious had I not embarked on my spiritual path?

Considering how bad the hangovers had gotten by 2010—the year I went on my first sober weekend getaway at the Ibiza yoga retreat, followed by my miraculous Monday morning moment—the writing was, without doubt, already on the wall.

I had been blaming "work stress" for the fact that anxiety had become my default emotional state—the kind of stress that makes you tearful and confused and robs you of your sleep, as you wade toward the end of each day through a quagmire of *hopelessness* spiked with electrical shocks of *fear*. With hindsight, I can also see that I'd begun to drink more and more alcohol to achieve the same "buzz" and sense of relief, resulting in more devastating and protracted bouts of "wine flu." And beer and vodka and Prosecco flu.

Floating into the office that fateful Monday, floored by how much calmer, more confident, and more *alive* I felt having abstained over the weekend, there was no more denying the link between my anxiety and my alcohol habit. If I wanted to feel calmer, more confident, and more alive in general, evidently the booze would have to go.

Oh. Crap.

Friday night drinks were often one of the only things in my week worth dusting off a smile for. And so, the question, *Would I have gotten Sober Curious had I not embarked on my spiritual path?*, in fact means, Would I have had the guts to bypass the easy escape hatch of a vodka martini (or three) and embark on my quest for a lasting fix had I not taken the opportunity being dangled in front of me to begin with some serious questioning of my life choices? To begin seeking some ways to address the past and unhealed hurts that my overall sense of malaise told me were festering just below the surface? Or would I have kept tumbling down the same old rabbit hole, winding up hungover, back at square one, for the remainder of my days?

People in AA talk about the problem with "dry drunks," meaning those who stop drinking but who otherwise do nothing to address the behaviors, habits, or emotions that led them to drinking in the first place. In *A Nice Girl Like Me*, Rosie Boycott writes of the miserable state of affairs that is dry drunk syndrome: "No inner development takes place at all; the lying is still there, so is the irritability, so is the tenseness and disharmony.

Working on the premise that life without booze must be better, people stop drinking and wait for their worlds to improve, as though they are owed a favor for having made the supreme sacrifice of cutting out the sauce."

Her implication is that this inner development, which could also be termed *spiritual self-inquiry*, is the real key to kicking the "sauce" once and for all, discovering the unanticipated joys of not drinking, and conquering FOMA for good.

Whatever your reason for getting Sober Curious, as Russell Brand writes in his 2017 sobriety memoir *Recovery: Freedom from Our Addictions*: "Because what we are dealing with is a spiritual condition—a post-religious spiritual calling—the inner condition is what we must discuss." Meaning, yes—once you've dealt with the (by no means trivial) practicalities of removing alcohol from your life and your relationships, moving forward is going to mean journeying *within* to reacquaint yourself with the person who chose to do all the boozing in the first place.

One thing to get clear on up front is that when I talk about "spirituality," I speak not of organized religion. Step Two of the famous AA 12 steps is "We came to believe that a Power greater than ourselves could restore us to sanity," and for me, that "Higher Power," which has helped get me to a place where I no longer desire to drink, is my own spirit, soul, or connection to Source. It could even be defined, if it helps for me to strip some of the "woo" out of it, as my essence or *life force*.

In fact, one thing that put me off AA was its religious overtones. Steps Three, Five, Six, and Eleven all make specific reference to the G-word (God), while Step Twelve sums things up thusly: "Having had a spiritual awakening as the result of these steps, we tried to carry this message to alcoholics, and to practice these principles in all our affairs." Enter the aforementioned "evangelical phase."

Yes, the clarity that comes from cutting out booze can feel like something of a "spiritual awakening" and being fully present with yourself is a fast track to connecting with whatever mystical force keeps pumping life into your being. For me, though, it happened the other way around. Some readers may know me as the founder of The Numinous, a cosmic lifestyle platform that updates all things New Age—for life in what I call the *Now Age*. Astrology, yoga, meditation, tarot . . . you name it, I'm into it, and I wrote about how I use these tools and practices in my first book, *Material Girl, Mystical World*. Some among you may also think this makes me sound like a total flake, and you are of course entitled to your opinion. Please rest assured that I am not going to try to "convert" you to anything in this chapter.

But I describe these as my *spiritual* practices because they are what help me connect to my spirit, or essential aliveness—and I would term anything that helps you do the same *your* spiritual practice. If now you're sitting there thinking, "Well a shot of tequila makes me feel more alive!"—my theories on that are wide-ranging, which is exactly what we'll be getting into in just a minute.

In any case, leaving my mainstream journalism career, launching The Numinous, and diving Alice in Wonderland-style into a world of psychics, planetary transits, and cosmic quests to the deeper realms of consciousness has been the backdrop to my Sober Curious awakening. In fact, me finding the missing pieces in my ongoing mission to reframing my relationship to alcohol became the surprise subplot of writing *Material Girl, Mystical World*, to the extent that the final section of the book is titled "People & Parties." This includes a chapter called "Healing Is the New Nightlife" that opens with a scene about the time I went to a breathwork circle in a teepee in Havemeyer Park in Brooklyn and got, like, MDMA high off *breathing*.

That kind of experience—accompanied by some soul-shaking realizations about myself, my life, our planet, and the kind of person I even came here to be—has been the lifeblood of my Numinous adventure. And fuel for my Sober Curiosity.

Not to mention the fact that I was moving in circles where alcohol was perceived very differently. A lot of my new friends worked in the healing arena, as Reiki masters, intuitives, and energy workers. And you can forget FOMA; these people were high on vibes. "I just don't really think about alcohol anymore" is something I heard A LOT.

But that was most definitely NOT the case for me. I might have been steadily drinking less and less by that stage, but I was still doing plenty of thinking about drinking. An obsessive amount, you could say, which was what led me to AA. When I spoke to an expert from The Cabin, a high-end rehab facility in Thailand, on the pretext of researching a story (basically me getting Sober Nosy), he told me that the obsessing is as much a part of the "disease" of addiction as the using. By this logic, the fact that I'd spend all week before a dinner date internally debating exactly how many glasses of wine it would be "okay" to drink clearly made me an alcoholic, whether I had the stuff or not.

Perhaps.

One thing was becoming clear as day, though. Which is that so many of the qualities I had always sought through consuming *spirits*—relaxation, a lift in my mood, connection, community, transcendence—were also being facilitated by a deepening connection to *my spirit.* In fact, with every gong bath, Kundalini yoga kriya, and energy healing session, I was becoming more and more convinced that alcohol had been a second-rate stand-in for the joy, inspiration, confidence, contentment, and overall sense of *aliveness* that I was being reminded I am perfectly equipped—not to mention *cosmically designed*—to generate for myself.

The Link Between Spirits and Spirituality

One of the theories for how the tequila kind of spirits came to be called that is that the vapor given off and collected during the distillation process was thought to be the spirit, or *essence*, of the original material. And ever since I got Sober Curious, I've been struck by the perfect irony of this—how imbibing the tequila kind of spirits also appears to mimic the effects of feeling connected to our own spirit, or sense of aliveness. Not to mention how we often turn to alcohol as a shortcut to feeling more *spirited* (lively, joyful, inspired) when life circumstances have crushed the life out of our own spirit.

Alcohol also plays a role in many organized religions—from the red wine used to symbolize the blood of Christ in Christian communion to the ceremonial blessing and imbibing of wine during the Jewish Passover. At the more far-out end of the spiritual spectrum, I've also met some mystics who believe that when you drink alcohol, your spirit renounces your body, creating room for other spirits (or non-true versions of yourself) to take over. Which would be one way to explain why we use the term "out of it" to describe acute drunkenness or why we use booze as a way to "escape" from our earthly problems. All those crazy, not very sexy or cool things you said while you were wasted?—that you also have no recollection of whatsoever? Maybe it wasn't even "you" saying those things at all, but a nasty little demon with a chip on its shoulder!

In the tarot deck of cards, meanwhile, which have been used for divination for several hundred years and which I think of as being like "Google for the soul," addictions and self-sabotaging behaviors are ruled by the Devil card, suggesting that the associated obsessions and compulsions are literally the human experience of "hell." (*Anyone?*)

For the more mystically minded, drunkenness as a gateway for dark entities to interact with us is one way to explain what happens when we black out—not exactly a scientifically sound theory, of course. The science of the blackout is fascinating in its own right, however: When exposed to alcohol, the brain receptors that create memories in the hippocampus (the part of the brain that's also responsible for emotional responses) shut down. Depending on the extent to which those receptors are disrupted, a blackout can be either partial (known as a "brownout") or complete ("en bloc"—aka "that never happened").

But whichever approach you prefer—knowing the "how" or pondering the mysterious "whys"—one thing is certain: When we're under the influence of alcohol, we're not really *all there.*

And if anything, I see embarking on any kind of spiritual journey as being about seeking the opposite—a sense of *wholeness*.

In a letter to the cofounder of AA, Bill Wilson, in 1961, Carl Jung, sometimes recognized as the father of modern analytical psychology, spoke to the need to include spirituality in the treatment for alcohol addiction by quoting the Latin phrase "*spiritus contra spiritum*." Commenting on the alcoholism of a former patient of his, he wrote, "His craving for alcohol was equivalent on a low level of the spiritual thirst of our being for wholeness, expressed in medieval language: the union with God. How could one formulate such an insight in a language that is not misunderstood in our [secular] days?" He ends the letter with: "You see, alcohol in Latin is '*spiritus*' and you use the same word for the highest religious experience as well as for the most depraving poison. The helpful formula therefore is: *spiritus contra spiritum*." Or, loosely, it takes spirit to cure spirits.

It could be said that we achieve this wholeness through living with integrity—a word that describes both the physical state of being whole, and the act of aligning every thought, word, and deed with what we know, in our heart of hearts, to be true (leading to what I think of as *cognitive resonance*, remember?). When we really begin paying attention to all the ways in which we bend our personal truths in order to fit in, be liked, make others happy, and not rock the boat, this kind of alignment is far easier said than done. Seriously, try noticing every time you bite your tongue, edit your words, or even judge and curtail your own thought processes for just one day, and you'll see what I mean.

Hence all the *spiritual tools* to keep us on track with the integrity. For me, this might mean my astrology practice, to help remind me of who I am underneath all the conditioning (familial, societal, generational, etc.); meditation and yoga, to train me to monitor my thoughts and reactions in real time; or the teachings of shamanism, to help me seek insights into the roots of my disorders so I can confront them at their source.

But wait, alcohol also gets called the "truth serum," doesn't it? A magic potion that helps us relax, drop the masks, and open up. *In vino, veritas*, and everything. When I asked my friend Sah D'Simone, a meditation coach who teaches about the links between science and spirituality, about this, he told me: "When we drink, we often feel like we're living our truth. We have shocked our nervous system with a whole new vocabulary, and for a few hours it's as if everything aligns. But the reality is, we don't really know how we acted, or what we said while we were drunk. The next day, nothing has been integrated. The parts of the brain that need to fire together to create a positive new memory or belief about yourself didn't go online at the same time. It's like cheating on an exam."

In a wider sense, the kind of spiritual path I'm talking about is a quest to truly know, accept, and integrate every part of our essential selves and our personal histories. Often going back generations! Parts we may feel ashamed of, confused or traumatized by, and that we have learned to edit and airbrush or else keep hidden to "keep the peace," both for ourselves and for others. What Carl Jung termed the "shadow" self, and our personal *demons*, you could say. And perhaps they are the same ones that come out (or swoop in) and talk all that shit when we're drunk. Parts of ourselves that are often desperate to be shown some love and acceptance. Given the ubiquity of alcohol, and our secular society's general acceptance of scientific over spiritual "solutions," alcohol may have been the only tool we've stumbled on for switching off our thinking brain in order for these parts to be expressed—even if often all drinking really does is numb us to the demons' screeching for attention, meaning, they'll be back with a vengeance the morning after since little has actually been done to address them.

As for Rosie Boycott's observation about dry drunks believing that they're owed some kind of cosmic favor for getting sober? If my own spiritual path has taught me anything, it's that there are no shortcuts and no favors when it comes to living with integrity; that the lasting and sustainable joy, confidence, and bliss that are the desired by-product of all this will find you only when you actively choose to seek them.

It's a journey that is by no means all "light and love." It's a journey that requires patience, courage, and serious grit. Remember how in *This Naked Mind* Annie Grace writes, "When we enjoy the 'pleasure' of a drink, we restore the wholeness and peace of mind we knew our entire lives before we ever drank a drop"? For the majority of us who find ourselves seeking this "wholeness" in drink or other substances, chances are there's something in our past that must be addressed, tended to, and lovingly reintegrated before we can ever again experience true "peace of mind."

Ideally, this happens with the support of a community of individuals who love us just for being us and who have our best interests at heart. And potentially with the guidance of a trained mental health professional. AA provides the community aspect for some people, and I've included a list of alternative resources at the end of the book. Because, if choosing to walk a spiritual path often means embarking on a journey to the murkiest, dirtiest depths of our souls, then it could potentially entail getting up close and personal with your shit-talking demons and confronting your fair share of devils. The Force is, of course, always with you. But it never hurts to get fully tooled-up.

The Real Genie in the Bottle

As Carl Jung noted, the irony is that drinking can often begin as a spiritual quest of sorts. I find it interesting that many popular spiritually minded authors have a background with alcohol abuse—Brené Brown, Cheryl Strayed, Glennon Doyle Melton, and Gabrielle Bernstein, to name a few. Did their spiritual seeking begin with booze, too?

As Russell Brand writes in *Recovery*: "The yearning itself [to drink or use drugs] is real, it's trying to lead you home." Meaning, as noted, that when truth, inspiration, joy, liberation, and aliveness are lacking in our lives, booze can be the first and the fastest way we find to fill in the gaps. Not to mention that it is sold to us as precisely this by a society based on consumption—a society that is also suffering what Brené Brown calls a "crisis of disconnection."

In fact, the abuse of alcohol and other drugs (including prescription pharmaceuticals, and not to mention social media) could be seen as a natural by-product of our consumer culture—a culture that is ingrained in us and tells us that amassing more and more things (including "friends" and "likes") is the fast track to happiness and contentment (*wholeness*, even), as we also strive to attain a certain level of status in the eyes of our peers. A degree of material security is necessary for our survival, yes, as is being seen as a valuable member of the tribe (and therefore worth including in the overall power structure); but our fear of not having or being "enough" has also been routinely exploited by the capitalist system, resulting in a neurotic need to consume more, and more, and more in order to feel safe, seen, and fulfilled.

Since consumer culture, or capitalism, also breeds competition (if I want something I'd better grab it before/earn more than the next person), one could argue that a constant quest to be "the best" is also at the heart of Brown's connection crisis—in itself, the fuel for many of our addictions. So says Johann Hari, whose TED Talk titled "Everything You Think You Know About Addiction Is Wrong" had been viewed almost 9.5 million times as of summer 2018. In it, Hari seeks to illustrate that "the opposite of addiction is connection," quoting the work of Peter Cohen, director of the Centre for Drug Research in Amsterdam. Hari says that Cohen emphasizes human beings' "natural and innate need to bond, and when we're happy and healthy, we'll bond and connect with each other, but if you can't do that because you're traumatized or isolated or beaten down by life, you will bond with something that will give you some sense of relief . . . That might be gambling, that might be pornography, that might be cocaine, that might be cannabis, but you will bond and connect with something because that is our nature."

When we're ashamed of our perceived "otherness" and have been programmed to believe that what we have and who we are is not "enough" (to make us lovable and valuable to our peers), alcohol, a substance that *switches off the part of the brain that monitors how we are perceived* and that also appears to be the lifeblood of our social connections, becomes very appealing indeed.

Buddhist philosophy, meanwhile, emphasizes that human nature is in fact *to suffer*, and this suffering (*dukkha*) is caused by attachment or desire (*tanhā*, literally, thirst)—including the desire to connect, even. This mirrors the neuroscientific view that all human behavior is motivated by the desire function of the brain, suggesting that addiction, too, could simply be regarded as another aspect of the Buddhist take on human nature—that both the spiritual and the scientific views are, as I said in chapter 1, *we're pretty much hardwired to get hooked on hooch.* "But there is a way out of suffering, and that is to liberate oneself from attachment," says Sah D'Simone. The path to nirvana—or enlightenment—is to extinguish the fires of misplaced desire.

Which sounds so simple, but which is in fact a lifelong practice.

Again, alcohol and other addictive substances can appear to facilitate this. How often have you heard the phrase "*drinking my cares away*"? And aren't the things we "care" about the things to which we're most "attached"? The things we maybe even "desire" the most? But as we know, this kind of liberation is fleeting and comes at a very high price—some might say, means literally making a deal with the devil.

Which is not to say *stop caring*! But rather, the way to end the cycle of suffering could lie in beginning to question *what* we're most attached to, and *why*. Which of our desires are rooted in greedy or destructive urges, and which are in service of our personal growth and the betterment of society? And we need to simultaneously question the belief that having certain things (from a romantic relationship with *X* person to a swanky new job or a parent who no longer judges our life choices) will make us "happy"—as opposed to our being happy just being *ourselves*.

The ways in which to begin this line of self-inquiry are manifold, and I have already mentioned some of my preferred tools. The first step, however, is to stop confusing matters by throwing more alcohol at the problem. Clarity is the goal here, people, the better to hear the voice of your soul—the (higher) Self, if you like, *who truly knows what's best for you*.

Russell Brand believes that our cravings, which usually manifest in me as restlessness, boredom, or an overabundance of manic energy, are actually this Self trying to communicate something to us. "There is something in you speaking to you and you don't understand it because you've never learned its language," he writes. "Spend time alone, write, pray, meditate. This is where we learn the language." Where we get to work out what we *truly care about*, at soul level, he means.

Marc Lewis told me that "the current evolution of self-care, self-help, and especially mindfulness meditation, it's great. Not only do [these tools] help people get past their addictions, in fact, I think these are probably the most powerful tools that anyone can use." I would go even further and suggest that the *reason* these tools are becoming so popular is that more and more of us are becoming disillusioned with the capitalist promise of fulfillment through consumption (including alcohol consumption) and are beginning to seek some answers of our own.

And all in the name of addressing the biggest question of them all . . .

Who Am I (Without Alcohol)?

I think a spiritual journey is not so much a journey of discovery. It's a journey of recovery. It's a journey of uncovering your own inner nature.

—BILLY CORGAN OF THE SMASHING PUMPKINS

If, ultimately, seeking a connection to our spiritual or higher Self is about seeking our true identity, then it makes sense that my Sober Curiosity was kicked into high gear when I moved from London to New York City. On paper, this looked pretty fabulous. People seemed either impressed or downright envious, as if somehow the Pisces and I had scored a special upgrade to platinum #LifeGoals status. A telltale sign of how *I* really felt about the move—which entailed leaving my family, friends, and my hard-won magazine career behind—was when I began crying into my pint when the Pisces and I hit the pub to celebrate.

Because I have since learned how my soul speaks to me, I now know the tears were its trying to warn me that the existential shit was about to hit the fan. Because in reality, everything I thought

I knew about who I was and what made me popular, successful, and cool was up for review. Of course, the reasons I drank were deeply entangled with how I saw myself. And since my being "Cocktail Girl" was also a huge part of my carefully constructed identity on the London scene, it makes total sense that the move would also ask me to investigate the role alcohol played in my life.

A lot of people come to their spiritual paths after a major life change and/or trauma. The death of a loved one. A divorce or devastating diagnosis. I feel extremely fortunate that my life "crisis" did not involve one of these. But along with being stripped of the status that came with my London media career, as I simultaneously began work on The Numinous and inched my way out of the spiritual closet (read: revealed my inner "woo"/weirdo to the world), my subsequent Sober Curiosity demanded that I journey back in time to integrate once and for all the damage done by my eating disorder and my abusive relationship with the Capricorn.

It truly wasn't until writing the intro to this book that I made the link. How it was alcohol—not friends, career success, or even meeting the Pisces—that helped me to move swiftly on from this period of my life, as I congratulated myself for having "gotten over it" so fast. How I actually created Cocktail Girl as a way to gloss over the shame. Shame about having been mentally, emotionally, and sexually abused from the age of sixteen to twenty-two. Shame for having simultaneously hurt my body by starving myself. And shame for having *chosen* to disempower and numb myself in this way over confronting the feelings of "otherness" that resulted from my unconventional and unstable family background. *What kind of a person does that to herself?*

Not a person who loves and values her Self—and is therefore demonstrating that she is worthy of being loved and valued by anyone else, that's for sure.

As if I had pressed "pause" on the feelings the relationship left me with the moment I walked out on the Capricorn and into my new Cocktail Girl persona, it all came flooding back as soon as I began to take extended breaks from drinking. Meaning that the Capricorn, who I had banished from my thoughts the day I left him, would literally appear to me in "psychic" visions on my yoga mat. That as sobriety led me to finally confront ongoing digestive issues, I began to reflect on the real reasons for my eating disorder. Things hit home when a mutual friend of ours reached out *sixteen years later* on Facebook, and seeing the Capricorn's picture in her photos caused a full-body jolt of adrenaline, as if I'd been punched in the stomach.

Obviously, I hadn't gotten over it at all.

Integrating the above meant first feeling all the shame. Then the sadness. And then the rage. I had to allow it all to flow out of my body on a gushing river of tears. And then fully own my own role in the situation, my having opted in to this sorry state of affairs, for whatever reason, as opposed to seeing myself as a victim. It was only when I finally found a way to forgive the Capricorn (through meditation, writing, tantric work, and astrology to help me understand my karmic wounds, for example), not to mention myself, that I was able to move on. And I thank Goddess for the people in my life—my mom, the Pisces, various healers, and a few close friends—who held and supported me on the journey.

I also came to recognize the Capricorn as one of my first addictions, since handing the reins over to him allowed me to "get out of" confronting my feelings about my parents' divorce and my insecurities about becoming a sexual woman. Anorexia was a way to leave my body and "escape" from *him*. And then finally alcohol stepped in. So pleasurable, glamorous, socially celebrated, and confidence-boosting by comparison! No wonder it stuck around the longest.

I share all this not to gain your sympathy, but to illustrate the dots that may begin to join up for you when you dig deeper into your personal boozestory. The kind of "stuff" that may show up demanding to be dealt with once the initial feel-good factor of getting Sober Curious wears off. (Which it will, by the way. Sorry, spoiler alert: Life may be a lot "better" and more joyful without the hangovers and the overall physical *funk* that comes from regularly putting a toxic poison in your body, but it definitely doesn't get any *easier*!)

It may not be something obvious or particularly tragic, but we all have pain lurking somewhere in our pasts. Even if the only evidence is the nagging feeling that something vital is missing, that we are not quite *whole*.

Could it be that what's missing is a clear and cohesive understanding of our own personal history? And that the spiritual seeking might be a drive to "recover" this? Marc Lewis profiles two addicts, Natalie and Brian, in *The Biology of Desire*, describing how they "began to outgrow their addictions [to heroin and crystal meth] when they were able to reflect on their lives, connect their past to their present conundrum, and imagine a future very different from the present." In other words, a future in which whatever pain that led them to use had finally been integrated and healed. In which the demons had been silenced once and for all.

Many healers past and present have used what can be called "story medicine." Through telling the stories of our own lives, we can acknowledge past traumas, the pain and the shame they have caused, and our fear that because of them we are in some way defective. Other. No longer worthy of love and connection. The "medicine" lies in sharing these stories with others, allowing us to be seen and accepted in our wholeness, shit-talking demons and all.

One of the best things about AA is the storytelling, which is largely what happens behind those closed doors. I believe the element of "confession"—beginning, I guess, with "admitting" that you have become powerless over alcohol—is the core of the program's success, even if the underlying message of "once an alcoholic, always an alcoholic" is a story that can keep us stuck in a negative feedback loop, since as much as we tell the stories we live, we also live the stories we tell ourselves.

I share Rosie Boycott's view on this: "There were many things in my history which I valued. If alcoholism is a spiritual search, and I certainly was coming to understand my drinking years this way, the last thing I wanted was to be told to forget my life up until the moment I walked through the doors of my first AA meeting . . . The seeker was well-intentioned . . . I was fond of her . . . and I had no intention of eradicating her from my life altogether."

Rather, the Sober Curious path is to welcome that seeker in. Into the home of your Self. To recognize your addictive or compulsive behaviors as the yearnings of your soul. To own and accept, without shame, loud and proud, the fullness of your story. And in doing so, begin to understand that it is also within your power to write yourself a brand-new script.

WINE IN THE AGE OF WELLNESS

Okay. We just went pretty out there (or rather, in there). Let's bring it back to basics: what booze is doing to your body. This is the chapter in which I'm supposed to list all the reasons alcohol is wrecking your health, in the hope of scaring you into not drinking. Which is not what I'm going to do.

First of all, I'm not a doctor or a scientist and so I would essentially just be Googling a bunch of stats to back up my own anecdotal investigations. Which, if you feel so inclined, you can go ahead and do on your own time. Some suggested search terms you may want to dive in with—if you want to try to scare *yourself* into not drinking, that is:

Alcohol linked to which chronic illnesses

Alcohol cause of what percentage of preventable deaths

Alcohol effect on major organs of the body

This is not my top recommendation since it obviously doesn't work. We all *know* that alcohol is bad for us and yet we all drink it anyway, often while pointing ("See! It's fine!") to the articles that appear on rotation that let us know drinking in moderation is *good* for our hearts. "Moderation" often being the key word in these kinds of missives. A word that, as already established, does not have a place in this book—since moderation also doesn't work. That is, if your goal is to radically shift your perspective on booze, get Sober Curious, and begin experiencing the life-changing benefits of not drinking.

If I've learned anything in my twenty-something years working as a lifestyle journalist, it's also that statistics—or at least how they're cited—are rarely totally objective. Depending on the case you're trying to make (that people getting Sober Curious is not only a growing trend, but could even be described as a *movement*, for example), enough Googling will generally produce the numbers to back up your argument.

There's the 2016 study that showed that the number of teetotal sixteen- to twenty-four-year-olds increased by over 40 percent between 2005 and 2013. This is confirmed by a report dated the same year from the UK Office for National Statistics which showed that the proportion of Britons overall who regularly drink alcohol had dropped to its lowest point since 2005.

And then you can do some more digging and argue the opposite.

In December 2017 it was reported that the British National Health Service is considering the use of "drunk tanks," sort of like hospital holding cells where people can sober up, to relieve the increasing pressure put on emergency rooms by intoxicated "revelers" (accounting for up to 70 percent of visits on Friday and Saturday nights).

In the US, meanwhile, a 2017 report published in *JAMA Psychiatry* showed that alcohol use is on the rise, up 11 percent overall between 2002 and 2013, and that high-risk drinking rose by 30 percent over the same period. These increases were most prevalent among minorities, women, seniors, and people with low levels of education and income.

So there you have (some of) the numbers, from which we could deduce that young people are drinking less, while risky drinking overall is rising. I have some theories of my own about why this might be, which I'll share in a moment, but the stats do not exactly back up my case for a mass uptick in Sober Curiosity.

One thing I can tell you *anecdotally*, however, is that ever since I started getting vocal about my own Sober Curious path, it's the subject I've been asked to write, speak, present, and podcast about the most. *What does it mean to be Sober Curious? Why did you decide to quit drinking? What are the biggest benefits? How has it affected your social life? How much do you drink now?*—these are just some of the questions journalists, bloggers, event organizers, and social commentators always ask, and which may be posed to you, too, as you walk this path for yourself. Not least because for many "normal" drinkers, the big question (*Would* my *life be better without booze, too?*) is often lurking somewhere just below the surface.

But the trend-hunters also want to know this: *Why is everybody getting Sober Curious* NOW?

Well, obviously, "everybody" is not. If anything, the numbers paint a different picture—one that is particularly troubling when we consider the rise of alcohol abuse among the most disenfranchised and marginalized. But from where I sit—watching the numbers of attendees at our Club SÖDA NYC events spike, seeing nonalcoholic "social tonics" and booze-free spirits pop up on cocktail menus, and being asked to speak to drink brands that are seeing a dip in revenues—our collective Sober Curiosity is beginning to feel like the murmurs of a *revolution* when it comes to our relationship to booze.

Which brings me to that "*why now?*" Why is it that the Sober Curious *conversation* is gaining traction—even if not everybody is on board with the actual not-drinking part of the equation (yet)? And why is this so much more than a fluffy lifestyle trend?

Well, first and foremost, let's take a look at the news, which will swiftly remind us that we are living through times of rapid, unprecedented, and often unsettling change. Advances in technology, globalization, and the destruction of our natural habitat, as well as an increasingly unstable economy, mean that our futures seem less and less certain. This very real uncertainty and unease (though sometimes presented as overblown fearmongering) about our futures is leading to record levels of depression and anxiety. One way (the educated way) to mitigate this? Step away from nervous system stimulants such as alcohol and seek, instead, calming and restorative ways to switch off from the headlines and unwind (see the mainstreamification of yoga and meditation).

Those teetotal sixteen- to twenty-four-year-olds? Having grown up with so much information and so many headlines at their fingertips on the internet, they are by far the most anxious and also far more likely to have educated themselves on the toxic effects of booze. (Also, to have established that being drunk in selfies is not a good look.) Not to mention that they have far more to worry about when it comes to the future of our planet. They're the ones who'll be living with the collective hangover of our consumer culture, after all. For younger generations there is the instinctive sense that this is no time to be getting "out of it." Real solutions to the real-world problems we're facing will require us to be very much *in it*.

Second, alcohol does not exactly fit with our culture's current and collective #WellnessGoals. Chances are, one of the reasons you picked up this book is because you are somewhat invested in your well-being. You are one of the millions of people who, in an uncertain world (with escalating doubts about the ethics and effectiveness of our health-care systems thrown in), are seeking a degree of security by taking their health into their own hands. Perhaps you've been experimenting with your diet; have gone periods without gluten, or dairy, and are contemplating (vegan ice cream not included) doing the same with sugar; have gotten hooked up to a fitness tracker. Maybe you've even joined the ranks of the "mindfulness" believers and already made yoga and meditation a part of your weekly routine.

And having dipped a toe in the practices that have helped make "Health and Wellness" worth three times more globally now than the pharmaceutical industry, perhaps the disconnect between the way you feel after your Thursday night vinyasa yoga class and your Friday night session on the vino has become more and more apparent. Is becoming harder to ignore. Maybe you're even feeling so good as a result of the time, energy, and money you're investing in your well-being that you no longer see the point of "cheat" days spent drinking all your good intentions down the toilet.

Especially when hangovers now come laced with an added layer of twenty-first-century existential unease.

Your Body Has Always Been Sober Curious

As I said, what you won't find in this chapter is a bunch of stats about how good or bad alcohol is for your health. And since when did we learn to trust the results of a Google search over the experience of our own bodies?

We don't need studies and statistics to prove to us that alcohol has no place in our wellness regimes because . . . HANGOVERS, PEOPLE! Headaches, nausea, vomiting. Fatigue, confusion, extreme thirst. Dizziness. Sensitivity to light and noise. The shakes. If you've ever had a case of severe food poisoning, chances are you've forever been put off eating what caused it. And yet being hungover—essentially nature's way of informing us that what we just put into our system *is poisoning us*—is simply seen as the inevitable payoff for another night on the sauce. One that can even be "cured" by getting back on it.

It's often only *after* the heart attack or the cancer scare that I've seen people willingly stop drinking for the sake of their health (and often not even then).

So, let's focus instead on the positives of not drinking.

For me, these have included more energy, dramatically improved sleep, better digestion, clear skin, a more optimistic outlook, increased productivity and confidence, greater self-acceptance, and a boost in my libido.

But don't take just my word for it. I also checked in with personal trainer Shona Vertue on this one (who is also Sober Curious, author of *The Vertue Method* [2017], and the woman responsible for getting David Beckham to do yoga), as cutting out booze is one of the first things she recommends to her clients. According to Shona, they report better brain function, fat loss, better muscle definition, improved recovery from workouts, glowing skin and fewer breakouts, reduced cellulite (particularly on the backs of the legs), and leaner tummies.

"So much can be achieved by just reducing alcohol!" she told me, while also confirming that drinking alcohol will "in no situation" result in their achieving their results. "Ever." As for alcohol and heart health? "The amount of red wine you'd have to ingest in order to reap the benefits of resveratrol (the compound in red wine that many claim has positive health benefits) would have you doing some very regrettable things before it helped your heart," she confirmed.

All of which is why getting Sober Curious is the next logical step in the wellness revolution. Not to mention that when you're feeling better as a result of the other diet and lifestyle changes you may have been experimenting with, the worse the hangovers will feel by comparison—and the more conflicting the inevitable aftereffects of drinking will be to write off as "the sign of a good night."

And yet, "when people begin working with me," Vertue says, "they're often prepared to do anything to get the body they want. Except give up booze [because] it's a very socially acceptable crutch and distraction. So much of our ability to socialize, find love [or sex], do business, and even relax [seems] alcohol dependent. It's very difficult to ask people to break free from that in the name of 'health.'"

Jen Batchelor, a wellness industry veteran and creator of the alcohol-free "conscious un-cocktail" company Kin Social Tonic, told me she was inspired to launch her business after witnessing "the hypocrisy of the LA yoga scene, where it was cocaine at 3 a.m. and yoga at 6 a.m. There was a disconnect between how people socialized and their wellness practices."

Reminds me of something my friend Mia, founder of the blog *The Sober Glow*, posted on Instagram recently: "Dear wellness community. If you are teaching people to live their 'best lives' but are not addressing the topic of alcohol consumption, you are completely missing the point." Said point being that alcohol is a toxin that is often used as a way to self-medicate deeper underlying issues, something that is rarely—if ever—discussed in wellness circles.

Reactions when I reposted this message ranged from, "Finally, somebody daring to mention the elephant in the yoga studio," to defensive arguments for "balance." But to me, the fact that we're so willing to overlook the mountains of evidence that alcohol is bad for us (let alone endure the hangovers) and instead focus on the "reported" benefits of the occasional glass of red, or argue that the stress we're trying to relieve with alcohol is worse for us than the drink itself, speaks volumes about our emotional attachments to booze—and suggests that more of us than would ever feel comfortable admitting are perhaps kind-of-just-a-little-bit-addicted-to-booze.

After all, "not a lot of people are sitting around and arguing that we should moderate cigarettes instead of just not fucking smoking," says Holly Whitaker from Tempest. "[Alcohol] messes with your endocrine system. It messes with your blood sugar balance. It does all of these things, and is a toxic, toxic, toxic, addictive drug. I'm like, 'Why are we fighting to keep this in our life and we're so ready to throw gluten out the door?'"

Hmmm, why indeed? Maybe because with gluten in particular, going "gluten free" (unless you are one of the approximately two million people in the United States who have celiac disease) has become a socially acceptable way of declining the pizza, pasta, and bread we still (also misguidedly) associate with weight gain—meaning a way that is less likely to get you, women in particular, labeled a narcissistic, fat-phobic, bad feminist killjoy.

We see this with sugar, too, which is the latest wellness bad guy, and so much cooler to decline for "health" reasons than because you're worried about your waistline. Although sugar addiction is certainly no joke—it being the leading culprit in the obesity/diabetes epidemic that's now recognized as the leading cause of preventable deaths in the United States (okay, okay, followed by tobacco and *then* alcohol, if you really want to know).

As noted, when I first got serious about reducing my booze intake, I was amazed by how many of my alcohol cravings were actually sugar cravings—a common occurrence, and one reason I don't recommend trying to quit both at once. Since alcohol has also been shown to increase the risk of all forms of dementia and many types of cancer (sorry couldn't help myself), you could consider it the greater of two evils and therefore worthy of addressing first.

Although if you do try my suggestion of switching up wine with dinner for a dessert instead, as you begin to navigate your *fabulous* new Sober Curious social life, you may also notice that you get sugar hangovers, too. The human body: It knows what it needs, and it knows how to let us know what it really, really doesn't need. Imagine what might happen if we chose to just shut up and listen for a minute.

All You're Really Addicted to Is Your Thoughts, and Meditation Can Help with That

Enter meditation—aka the holy grail of the modern wellness movement. I take it you have gotten the memo by now, how this ancient practice for quieting the machinations of the mind is being hailed as the antidote to everything from "stress" to insomnia and chronic pain. In fact, if one more person tells you to try to establish a regular practice, you might just scream. Because maybe you have tried and have found it surprisingly difficult to sit and focus on your breathing for ten minutes, or five, or even one! Unexpectedly challenging, unblissful, and therefore easy to leave off your to-do list.

If this is you, I apologize in advance, but I'm here to add to the chorus. This is precisely how we "just shut up and listen for a minute"—and when it comes to quitting an addiction, rewiring habitual patterns, and examining self-sabotaging behaviors (or whatever it is you've decided to call what you're doing by getting Sober Curious), all roads eventually lead to meditation.

After all, the fact that it can feel so excruciating to *just sit* is intimately linked to many of the reasons we use alcohol and other substances in the first place, whether or not we wind up doing so compulsively. As Tommy Rosen, author of *Recovery 2.0: Move Beyond Addiction and Upgrade Your Life* (2014), told me: "Most people think, 'Oh the habit is drugs, alcohol, food.' But, actually, the habit is avoiding the present moment. Meditation will ask you to be still and to be quiet. It is the immediate antidote. You could say the state of meditation is the opposite of the state of addiction."

The reason it can be so frustrating to try to meditate—*the thoughts, they WILL NOT STOP!!!*—also gives an insight into what makes the *distraction* of a drink (or an Instagram binge, a line of coke, or an online shopping marathon, for that matter) so very appealing. Because there is no escaping our thoughts. The nature of the mind is to think. And thanks to what Sah D'Simone calls our "negativity bias"—the fact that it takes five positive experiences to offset one negative one, an outmoded evolutionary trait designed to keep us out of harm's way—the most anxious, depressive, and unhappy thoughts are the most difficult to quiet of all. Enter the blessed relief of a readily available and swift-to-administer panacea, one that also creates the veneer of happiness-boosting social connection, at the end of a particularly negatively biased day.

In other words, no wonder we leap at any opportunity to get "out of our heads." And now I'm telling you that the way to freedom, joy, and ultimate peace of mind is to dive in there even *deeper*?

Yes. For it is this inner conflict between us and our thoughts that creates the craving. "It doesn't feel good to be put upon by our thinking, negative thinking especially," Rosen says. "So, naturally it'll place us into a state of dis-ease. From that state of dis-ease, we start to look for things [outside of us] to fix it." The actual "fix" comes from seeking to make peace with the constant chatter; it lies in learning to listen, without judgment, and to befriend what can feel like the absolute chaos of a confused, tired, and wired twenty-first-century mind.

Which doesn't sound particularly joyful, granted. Part of the reason being, according to Rosen, that "you're not used to being with yourself. You're not used to being with this present eternal now. You're used to avoiding it. And so, there's a great discomfort at first in just sitting and looking and not reacting, not doing anything." Particularly in a capitalist culture that tells us that the "solution" can always be found in doing or consuming another *thing*.

Yoga, another ancient practice co-opted by the modern wellness movement, offers a physical metaphor for exactly what Rosen is describing. Holding a difficult pose can be a reminder that whatever discomfort we may be feeling in the moment (ranging from boredom and frustration to physical pain) will, eventually, pass.

As with anything, meditation gets easier with practice—like, that *daily, regular practice* you keep hearing how you really should be cultivating. With practice, the negative thoughts that seem so overwhelming may not go away, but their immaterial nature becomes more familiar. They become easier to recognize as simply thoughts—often attachments or anxieties designed to help us either re-create past pleasures or avoid future pain—which it is within our power to choose to bring into the reality of the present moment. Or not.

"Choice" being the key piece to pay attention to here.

When I finally took myself to AA, having already reached a point where I was barely drinking at all, it was because of the random occasion when it felt like I had no choice but to get wasted. Not because of peer pressure or because the cravings had gotten *that bad*, but because waking up the next day, hungover, and wondering *How the hell did I let that happen?*, I felt like "I" hadn't had much say in the matter at all.

The "I" I'm talking about here could be described as the (higher) Self we met in the previous chapter. The "Higher Power," even, that they talk about in AA. Or simply the part of you that is able to "rise above" stressful situations, to "keep calm and carry on." A part of yourself that you come to know intimately when you begin to meditate.

Unfamiliar? Why don't I hook you up right now.

After reading this paragraph, I want you to close your eyes and focus your attention on your breathing. Set the alarm on your phone for three minutes or so, and begin breathing in for four counts, and out for at least five. Simply follow the breath in through your nose, feeling it tickle your nostrils and then expand into your body, and then watch it leave again. Repeat. During this process, your mind will wander. When it does, don't beat yourself up for being a bad meditator. Simply notice that you got distracted and come back to your breath. Clue: The noticing *is* the meditation.

Okay, go.

. . .

Okay, good! Were you able to focus on your breath? And did you notice how you got distracted by your thoughts? There are many ways to meditate, but this is one of the simplest techniques to experiment with. You are now qualified to do this any time you like, and you're welcome.

And now ask yourself: *Who, or what, was doing the noticing?* Your (higher) Self, is who. The part of you that some might call spirit—that is, the part of the Higher Power that connects all living things and that lives inside of *you*. Also, the part that *always* makes choices that align with your ultimate health, well-being, and (highest) good. Choices that are as unique to you and your needs as a computer password—including choices about what thoughts to focus on; how to spend your social life; and what food, drink, and other substances to put inside your body.

It's not a part we've necessarily been trained to pay attention to, especially given that these unique personal choices may often go against the status quo and lead us *away* from the messaging of our consumer culture. Not to mention that it's not nearly as shouty as the depressive, anxious, and unhappy thoughts our mind manufactures in a (valiant yet misguided) attempt to help us either seek pleasure or avoid pain.

If you want to get familiar with this part of you, to strengthen your ability to always make the "healthy" choice, then make meditation priority number one when it comes to your wellness regime. You can drink all the green juice, dodge all the gluten, and run around town in all the fancy yoga pants you like—but it means nothing if you're not also addressing your mental, emotional, and spiritual well-being.

Speaking of which, if the commodification of "wellness" is beginning to leave a nasty taste in your mouth—the industry as a whole often being criticized for marketing exclusively to wealthy white people and therefore to be unavailable to the individuals who may benefit from wellness practices the most—the best thing about meditation is, it's FREE.

The Blissful Sleep Piece

You know what else costs nothing and is about one of the best things for your mental, emotional, *and* physical well-being? A good night's sleep.

It's just coming up on 6 a.m. as I begin writing this section. Yes, I am a morning person—but not usually *this* morning, as I'm actually in what I call "jetlag training" ahead of an upcoming work and family trip from NYC to the UK. A week before I make the journey I begin setting my alarm ten to twenty minutes earlier each day, until I'm waking between 4:30 and 5:00 a.m. The idea (and it really works) is to mitigate the impact of the transatlantic time difference. Because you know what's the worst? Being shocked out of deep, middle-of-the-night sleep by your alarm going off five hours ahead of schedule—and it being time to get up.

For me, this is usually accompanied by a foggy head, an aching, empty feeling in my chest, and a creeping sense of dread/extreme grumpiness. And knowing what I know now about how alcohol affects our sleep patterns, when I look back at my drinking years it's like I was permanently jet-lagged—the sensation of the very worst jet-lag being not entirely dissimilar to the grogginess of a lingering three-day hangover.

Alcohol is one of the most common self-administered sleep aids, with an estimated 20 percent of Americans using booze to help them nod off. But as you will discover as soon as you remove it from your life, overall it has the opposite effect on sleep. I mentioned "dramatically improved sleep" as one of the main benefits for me of not drinking, and I am actually going to go right ahead and upgrade this one to MY VERY FAVORITE THING ABOUT NOT DRINKING.

Because *sleep, people.*

I have a friend who describes the velvety blackout sleep that the human body uses to restore energy, repair muscles, and reset our hormonal systems as "orgasmic sleep." It's the kind of sleep where you wake up feeling so deeply *satisfied*, your physical and psychic energies so replenished, that you can't wipe the smile off your face. The kind of sleep your body is letting you know it likes *a lot*. And the kind of sleep I get a lot of nights since getting Sober Curious.

To any new parents out there, I apologize if this comes off as annoyingly smug. But to the same new (and not so new) parents who may be using wine as a swift way to transition from kid-mode into me-time at the end of another *tantraumatic* day, know that *any* amount of Sober Curious sleep is better than an entire grown-ups-only weekend's worth of mildly sozzled sleep.

Why? It's all about our "sleep architecture"—or the basic structure of a night's sleep. Normally, we cycle between two types of sleep: non-rapid-eye-movement (NREM) sleep and rapid-eye-movement (REM) sleep. The sedative effects of booze make it easier to slip into NREM sleep, but REM sleep—when our muscles are completely paralyzed as we roam through the vivid dreamscape of our subconscious—is the most restorative kind of sleep, and the kind affected the most severely when we go to bed boozed up. Meaning that the sleep we do get—if we get any—is less than restorative.

I'm no stranger to the misery of insomnia, having experienced in my midthirties a bout of extreme "adrenal fatigue"—when the adrenal glands aren't functioning up to par, and symptoms such as debilitating fatigue, sleep disturbances, body aches, and digestive problems can appear. Beginning my day anxious and wired, to the point where I was often tearful, and then ready to fall asleep at my desk around 4 p.m. (when I'd reach for sugar in any format available to perk me up), I would crawl into bed exhausted—only to wake again at 3 a.m., monkey mind racing, heart pounding in my chest.

If this cycle sounds familiar to you, it's because adrenal fatigue is said to affect up to 80 percent of Americans; many alternative health practitioners consider it to be a by-product of "modern life." Among other factors, part of how it works is that any time we're exposed to a "stressor" (and this could be anything from work deadlines to relationships woes to the street noise and light pollution of living in a major city), our adrenal glands release the "fight or flight" hormone cortisol. Over time, we end up running on the stuff, feeling constantly primed to either fight or flee, as the adrenal glands become "fatigued"—leading to a variety of symptoms.

When I was "diagnosed" with adrenal fatigue, I was advised to cut coffee and sugar out of my diet because the stimulating effects of both exacerbate the viscous cycle. Which I did, which led to an immediate improvement, and which I swiftly became "evangelical" about. "Sugar is as bad as *cocaine*!" I told anyone and everyone, sounding annoyingly smug about my newly improved energy levels and mood.

But not once did anyone question my alcohol consumption (that old elephant again), and on nights when I had just a couple of drinks, as opposed to knocking myself unconscious with a full-on binge, the 3 a.m. waking was back with a vengeance. Not that I was willing to make the connection at the time.

I've since learned that this early waking is partly due to the sedative effect of the alcohol wearing off. There's the fact that alcohol, too, when consumed continuously over a period of time, has been shown to increase the production of cortisol. As for the jolt of adrenaline the body administers as your blood sugar level drops—this drop could be because the liver, which usually regulates blood sugar levels by converting carbs to glucose, is otherwise engaged removing toxic poison from your system.

Then there's the way alcohol makes you feel hotter as blood moves from your core closer to your skin, when cooling off is what guides us into the deepest part of the sleep cycle; the links between alcohol and sleep apnea (snoring and other forms of interrupted breathing during sleep); the diuretic effect, meaning multiple mid-sleep visits to the bathroom; and the increased chance of sleepwalking.

Oof. Tired of this topic yet? Got the message that booze will mess with your sleep as badly as a bawling baby? Great. Now what about all the good things associated with getting the regular, restorative, *orgasmic* sleep that you get to enjoy when you get Sober Curious? According to the US Department of Health and Human Services, routine deep sleep leads to:

A healthier immune system

Reduced stress

Improved mood

Better cognitive function

Other people seeming less annoying (lol—HHS actually included this)

Better decision-making abilities

Improved leadership skills

Fewer wrinkles and a glowing complexion

Bright, nonpuffy eyes (okay, I added this one)

Fewer cravings for salty and sugary foods (including booze! Win-win!)

All of which are similar to the benefits of meditation—unsurprising, since, according to Bob Roth of the David Lynch Foundation for example, "for many people, twenty minutes of transcendental meditation is said to be like two hours of sleep."

And then there's the less-immediate and therefore not-so-motivating but actually most-important fact that lack of sleep is shown to have links to chronic conditions such as obesity, dementia, Alzheimer's disease, and heart disease.

But we're not doing scaremongering in this chapter, right? Feel free to focus on getting curious about the concept of orgasmic sleep as your chief takeaway from this section as we move on to take a look at the bigger picture.

#Wellness as a Call to Something Bigger

The EtG urine test picks up traces of ethyl glucuronide, a byproduct of booze, in your pee as many as three or four days after your last drink, or about eighty hours after the liver has metabolized the alcohol. Imagine businesses performing random alcohol tests on employees the way some perform random drug testing. Would you pass?

In my drinking days, I most definitely would not have. I always made sure I had a few nights a week off the sauce, but three days was about the longest I'd ever go before topping up my urine alcohol levels with a cheeky midweek pint, or glass or three of wine. By this logic, from the ages of roughly twenty-two to thirty-eight, it's possible I never experienced a period when the effects of alcohol had completely left my system. And if it takes three to four days to pass a pee test, anecdotally I would say it takes about three weeks for me to feel the full, spine-tingling, life-is-fucking-beautiful joy of not drinking—for the aftereffects of as much as one or two drinks to vacate my mind, body, *and* soul, that is.

You know who gets to experience the delicious side effects of being completely alcohol-free, often without even questioning it? Ladies with babies. I've always been amazed how even the most dedicated drinkers seem to get Sober Curious overnight as soon as they become pregnant (or begin trying in earnest). This is most definitely not as simple as it looks for some—I've also met women for whom the challenge and misery of going nine months without a drink is what led them to question the nature of their dependence on booze. But in many cases, as soon as a woman's perception of her body switches from "mine to use and abuse as I see fit" to "sacred vessel for birthing new life," whether to keep pumping it with lady petrol becomes a no-brainer.

"I'm surprised how much I'm enjoying it."

"Everything feels less stressful."

"I just feel better in my head."

"I don't know if I'll even start drinking again."

These are just some of the comments I've received from pregnant friends experiencing the bonus, it could be said "glow-inducing," side effects of not drinking. The fact that the majority return to drinking once the baby is born and breast milk can be expressed—allowing Mom a night out on the tiles to feel like her "old self"—is a testament to how ingrained the belief is that *alcohol* is what makes everything in life "less stressful." Common wisdom about the dangers of drinking during pregnancy also means that a woman may not feel she has any *choice* in the matter, and since a woman's physical autonomy decreases rapidly both during pregnancy and after becoming a mother, the return of wine o'clock can also be read as a woman's understandable attempt to reclaim ownership of her body.

As for how this applies to you and your Sober Curiosity?

What if, as you seek the motivation and confidence it takes to walk this path, you imagined *you* were pregnant? Stick with me on this one! What I mean is, what if you—men, women, and readers of every gender expression in between—were to adopt the mindset that your body, too, is *a vessel for birthing new life*? Be it actual new humans or new creative projects, or positive new experiences and memories, or new ways to feel pleasure and transcend the day-to-day. Even new ways of thinking and being in the world?

Now about we zoom right out and imagine ourselves as collectively birthing *a whole new Earth*?

If the current wellness revolution is in part being fueled by fears for the future of our planet, one way to frame our collective Sober Curiosity could be that more and more of us are feeling called to the role of co-creators—the responsible and loving parents, if you like—of a future we can all believe in. One benefit of getting Sober Curious might be finding the time to launch a creative side-hustle, and another might be feeling and expressing newfound optimism for what the future holds. Or cultivating the self-belief that just maybe—rather than our being hopelessly underqualified for the job, and the task being too overwhelming to contemplate anyway—we can each play an active role in creating a happier and healthier future for us all.

And as for what "#wellness" has to do with *this*?

I actually don't like that word as it is used today because it has become shorthand for yet more commodification of something that is ours by birthright. Because "well" is how we (most of us, that is) came into the world. The external conditions we then find ourselves presented with—from our family history, to our social and economic status, to the education we receive—often are the factors that determine the state of well-being we experience overall in our lives.

As Biet Simkin defines it, "wellness" is truly "a platform on which to be a conscious creator. We've moved away from consumerism and into an era of experience being more valuable than commodities. And to experience something fully, you need to be 'well' to do it." Including to experience the fullness of our human potential.

Experiencing something fully also requires us to be fully present—something else we must consciously choose for ourselves in a world where our attention is the most valuable commodity of all, with online giants like Google and Facebook concocting endless ways to keep us fixated on our screens for as long as possible in the name of advertising revenues. If we're spending most of our days online or up in the cloud and our downtime downing shots, when do we ever truly get to *Be Here Now*? When do we get to choose what we genuinely want to experience next, versus what we're being sold? When do we get to make decisions and take actions that align with our (personal and collective) highest Self?

If our getting Sober Curious *is* the natural evolution of the wellness revolution, is on the brink of becoming a movement, it could be said that things are beginning to look up regarding the possibilities of where we go from here.

Some Yoga to Help You Get Sober Curious at Home

You don't have to go to a pricey class to reap the benefits of a yoga practice. I practice at home using a service called YogaGlo.com ($18 a month) that provides an online library of thousands of classes. Some of my favorites are with a teacher named Stephanie Snyder, who's been Sober Sober for twenty years and teaches what she calls "Yoga for Recovery." "In yoga, we say you can master the mind or the mind will master you," she told me. Here are some of her favorite poses and techniques for when the cravings/demons/repetitive old stories just won't quit:

- BREATHWORK. "Controlling the length and quality of our in and out breaths helps to balance the nervous system, and if the nervous system is integrated, well, we're able to process better."
- TWISTS. "Sending the top half of your body in a different direction to the bottom half helps to drive you into the midline, which is very, very, very effective at grounding and stabilizing your vibration."
- BALANCING POSES. "When you're trying to balance on one leg, for example, while doing something funky with your arms, you've got no other problems, right? You're just trying to hold yourself up, and so you get some relief from your thoughts in those moments."
- HEADSTANDS AND INVERSIONS. "Any pose that involves getting upside down brings blood to the brain and can be very empowering and refreshing."

- SHOULDER STAND. "Balancing on your shoulders with your legs up in the air or up against a wall helps regulate the thyroid and parathyroid, and includes a 'chin lock'—by pushing your chin into your chest—which is very sedating. So, it's very calming to the nervous system."
- HIP-OPENERS. "Our hips are like the junk drawer of the body. When we have an experience and we're not sure what to do with it—we might need it later, but we're not sure—we'll just shove it in the junk drawer. Our hips hold a lot of old emotional programming, and poses where we stretch this part of the body out help to release this."

CORE DESIRED FEELINGS

Unexpectedly *JOYFUL* and suffused with bliss. This is an alternative way of experiencing life that I have discovered on the other side of getting Sober Curious—much to my surprise, considering some of the darker places that this journey has taken me. Can I really guarantee that you'll make it here, too?

It's a pretty big promise, and one that's based on my personal experience, of course. But "joy" may mean different things to different people. Depending on all kinds of factors—family conditioning, education, social and economic backgrounds, and basic personal preferences, for example—the situations we term "joyful" will be varied.

That said, since I actually find sobbing jagged remnants of ancient pain out of my body to be, overall, pretty "joyful" (taking the accompanying feelings of release, acceptance, and liberation into account), please don't hold it against me if this is not the first word you'd use to describe your own Sober Curious journey! Depending on how far along you are by now, however, I hope that you are experiencing some of the wonderful new feelings that are generally the by-products of removing a heavy toxic load from your system.

For example, the toasty warm sense of physical well-being that pervades you as the poisoning effects of alcohol wear off. The relief of never waking up hungover. The sense of calm that begins to become the new normal—the cumulative effect of getting night after night of orgasmic sleep. The renewed sense of optimism and "I got this" that are a commonly reported side effect of removing a known *depressant* from your diet.

These are the obvious ones. And for anybody invested in their overall well-being, and who wants to boost their productivity, to experience more energy, and to feel better equipped to handle life in general, such side effects provide ample reasons alone to experiment with bouts of Sober Curiosity. But then, considering the overall toxicity of booze, you'd pretty much expect to feel some physical and mental benefits when you quit.

But the expanding sense of *joy* I hope you'll also experience speaks to something deeper.

As mentioned, after I originally got Sober Curious, I often found myself skipping down the street with *glee*—for no other reason than how good it felt to be *alive*. The irrepressible feelings of joy behind this kidlike behavior, which would also sometimes spill out as fits of uncontrollable laughter, usually kicked in only after at least three weeks of being booze-free—as if that's how long it took for the dregs of my last "drunk" to *completely* leave my system.

And part of me (the part, I suspect, that tends to spend the majority of its time manufacturing negative thoughts in a misguided-yet-valiant attempt to keep me safe from harm) would be standing there, confused, going, *"Who the HELL is she?!"*

Because I wasn't used to feeling that way naturally. At some point (possibly, considering my personal boozestory, at around the time I broke up with the Capricorn), I'd learned to *outsource* my experience of fun, relaxation, merriment, and, yes, joy to booze. My *happiness*, if you like. And in doing so, I had forgotten that all these "positive" feelings are as much a natural part of me as the "negative" feelings that often accompany all those negative thoughts. Considering the cultural conditioning that tells us alcohol is the fast track to fun, relaxation, merriment, and joy, and given its overall ubiquity and highly addictive nature, I suspect this is the case for many of us. After all, when it comes to feeling "good," why wouldn't we choose the fastest and most readily available route?

During my last experiment/reminder about drinking, I was awed by how just a few sips of beer could completely change my mood. When you don't drink, you tend to forget how freaking potent the stuff is! In an instant, I'd gone from mildly tense but hopeful for a positive outcome (my general resting state) to giving very few fucks, ready to kick back and relax with a big fat smile on my face. It was as if I'd slipped, almost imperceptibly, into a parallel reality in which everything was *golden*. And I hadn't even been feeling particularly *unhappy* in the first place. Of course, this reality lasted for only twenty minutes or so, before my recollections of the experience disappear into an echoing black hole. You know how it goes.

But how *very* seductive. The ability to switch off "stress" (our catchall term for those unwanted feelings) in an *instant*. Tumbling like Alice ("Drink Me") down a rabbit hole into a Wonderland of bliss.

For a regular ol' social drinker, whose habit isn't making them a danger to themselves or others, why worry about using this tried-and-tested formula once in a while? As one friend (a favorite drinking buddy at the time, who has stuck by my side and with whom I still have a total blast sober) said to me after I had told her about my early desire to quit drinking: "But you only drink to have fun. What's wrong with wanting more fun?"

Well nothing whatsoever! In fact, having more fun is one of my ongoing #LifeGoals.

But the danger of becoming a "risky drinker," and all that may entail, comes along when we confuse "drinking to have more fun" with "drinking to feel less pain." Given the muddling effects of the substance in question, not to mention the fluctuating, *numinous* nature of our differing feeling states, this distinction can prove very difficult to pinpoint.

How am I supposed to know where drinking to "feel good" ends and drinking to stop "feeling bad" begins? Aren't they generally two sides of the same coin? Take away one, and you're left with the other. And how did we come to label some feelings "good" and some feelings "bad," anyway?—some worth pursuing and some that require suppressing by any means possible (ideally, the fastest and most readily available)?

Enter the next, slightly more complex, set of questions to be addressed by our Sober Curiosity:

Will I ever feel happy and relaxed without alcohol? How will I switch off stress? Is my life so bereft of naturally occurring joy that the hangovers are worth it for the highs?

Happiness Is Our Natural State

Given our negativity bias, and the way our fears and anxieties are regularly exploited by the mainstream advertising industry to feed the capitalist system, it sounds too good to be true. But I would like you to remind yourself of this seemingly improbable claim throughout this chapter. Not many New Yorkers, for example, would agree with this statement—that Happiness Is Our Natural State. In a 2018 survey by *Time Out* magazine, 77 percent of people interviewed living in NYC said they are "stressed," and 91 percent said they regularly drink alcohol—meaning it could be assumed that a large majority of the stressed-out New Yorkers are also using booze to alleviate the pressures of living in one of the most expensive, competitive, and fast-paced cities in the world.

In fact, drinking to "relax" (i.e., switch off stress) has got to be one of the top reasons people everywhere use alcohol—even if, just as what brings me joy may be different for you, what creates the stress we are seeking to relieve is different for everyone.

But there are common themes. Perhaps money worries. Work pressure. Difficult family dynamics. Loneliness. A physical condition. Granted, we may not always have control over these external stressors, particularly those rooted in more intractable institutionalized sources of stress, such as systemic racism. For those in the most marginalized communities, for example, communities in which high-risk drinking is increasing most dramatically, life can be a daily struggle for survival. But when life gets tough, one thing we can do is not add to the load with a steady drip of not-so-awesome water.

And we can also learn to manage how we respond to said stressors. Which is where your meditation practice comes in, helping to create some room for negotiation between you (your all-knowing essential Self) and these pressures. A sense of perspective. A work deadline or a painful argument may be very real, but ultimately, they are not part of us. The extent to which we attach negative meaning to these experiences, our perception of them, is the extent to which we allow them to cause us pain.

I acknowledge that again, I'm speaking from a place of privilege, having never, thanks to the color of my skin and the circumstances of my upbringing, had to endure the daily, unavoidable oppressions of ingrained racial and social prejudices, for example. The situations in which my "otherness" put me on the outside having been few and far between.

The idea of shifting your perspective becomes far simpler when life has handed you multiple options on a platter, and under no circumstances will that shift make the stressors disappear overnight. We have tequila for that—even if we've learned by now that pressing "pause" on the pain, the fear, and the lack of self-worth for a few hours means they will only return with *a vengeance* in the morning—when it can feel like the demons are out for revenge for our willfully trying to suppress them. Like we have woken up back in our own personal hell.

As noted, "stress" can be a catchall for a wide variety of "negative" emotions—none of which like to be ignored, since all our feelings see themselves as having a valuable role in facilitating our overall well-being. In his book on emotional healing, *Senses of the Soul: Emotional Therapy for Strength, Healing and Guidance* (2012), spiritual teacher GuruMeher Khalsa describes our "heavier" emotions as "misunderstood friends, with benefits." He goes on: "Emotions are an essential part of our very accurate sensory system. They are a source of information to avoid trouble, heal the past, and help access the peace and happiness that are the goal of every being."

Peace and happiness which it could be said are not only the goal, but the baseline, default state of being that we naturally return to—once whatever led to anxiety, anger, or sadness stepping in has been resolved. You may well have experienced the relief, and *lightness of being*, that's the result of instigating a difficult yet necessary conversation or quitting a job or relationship that's been crushing your spirit. Could it be that *joy* was there all along, like a balloon held underwater always trying to bob to the surface?

Since alcohol is a known depressant, it makes sense that the immediate aftereffects of quitting drinking may include some buoyant skipping down of streets and eruptions of laughter. But once the initial *bounce-back* has passed, our newfound clarity will likely lead us to dig deeper into and address the root causes of our anxiety, anger, sadness, etc. At which point, a blissful sense of liberation can give way to what feels like some pretty heavy lifting.

Heavy lifting that I do NOT recommend you attempt on your own. If "a problem shared is a problem halved," then the community aspect of AA has got to be one of the most valuable of the program. In no other area of life is the same level of peer-to-peer support and counseling freely, and globally, available. And while the program didn't speak to me, you have absolutely nothing to lose in checking it out for yourself. I've also discovered since getting Sober Curious that there are all kinds of other support groups out there for quitting booze, some of which I list in the back of this book.

Depending on the nature of the stressors in your life, and the extent to which they are causing you pain, this may also be the point you choose to enlist some professional help. In the case of anybody navigating the deeper waters of emotional trauma (the lasting mark of painful or shocking life events), this level of support is invaluable, even essential—and I also list some organizations that offer counseling to people in need of financial aid. Or it may be that finding a Sober Curious ally or allies to confide in (as I did when I held the first Club SÖDA NYC meetup in my living room) offers support enough.

Along the way, though, be sure to notice, appreciate, and give *gratitude* for the times you *do* experience joy. The G-word (gratitude) is key, since it's been scientifically shown that recognizing all you have to be thankful for—even during times of extreme stress—helps foster mental and emotional resilience, improve self-esteem, and boost empathy. This means really sinking into the moments, with a giggle, when joy bobs to the surface of your being and you're reminded that it was there all along. What they used to call *stopping to smell the roses.* Choosing to trust that joy is *always* there cheering you on.

Not that this comes naturally to most of us. Our negativity bias means that we're predisposed to pay more attention to our heavier emotions and that we tend to mistrust displays of happiness and levity in others. It's easy for us to label people who appear to have accessed this mythical, ever-present internal state as "smug," "self-satisfied," or even "deluded." Not to mention how inbred puritanical attitudes about pleasure mean we may feel guilty for even experiencing the more pleasurable feeling states—subconsciously being quick to disown them for fear of appearing overly pleased with ourselves or for inciting the envy of others.

Ever bonded with your friends about how *miserably* hungover you all are? When everyone all around us is feeling like crap, a surefire way to fit in and be liked is to make sure we're feeling like crap, too.

And so, allow me to repeat. Feeling "good" is your birthright. No magic potion is needed to help you feel what is already in you—just a willing commitment to trust, appreciate, and embrace joy.

It's Time to Stop Taking the Edge Off

Which means it's time to expand on another of my key Sober Curious observations with you—namely, that choosing to walk this path can often mean learning to *get comfortable with being uncomfortable*. This practice, if not exactly joyful sounding, is where the real gems of this journey are often to be mined.

The degree to which this idea makes you, well, *uncomfortable* is a clue to how deeply discomfort-averse our society has become. When we're not medicating the first sign of any physical aches and pains out of existence, we're *seamlessly* ordering takeout and shopping from the squishy depths of our own sofas. Where consumer culture meets modern technology, our every possible need or want can be met at the *touch of a button*. Instant gratification. Which is pretty much anathema to building rock-solid emotional and spiritual muscle.

Have you ever watched a snake shed its skin? This cyclical process begins with it finding a sharp rock, or *edge*, to rub itself against, creating a tear in the old skin—which it then eases itself out of, inch-by-inch, to reveal a shiny new skin underneath. And how many times do you drink simply to *take the edge off*? Breaking that habit and changing your drinking for good (not to mention your *thinking* about your drinking) means making like the snake and confronting some of your personal edges head-on—ultimately so that the shiny new you can be revealed!

In my experience, these may include any and all of your Sober Firsts; not knowing what to do with yourself on Friday nights; being the only nondrinker on a boozy bachelorette weekend; cringing your way through dry networking events; navigating awkward friend break-ups; being faced with the rampant workaholism you drank to escape from. The "edges" (shedding/growth opportunities) therein are the uncomfortable feelings and jarring moments of self-reflection that are often the backdrop to situations like these, raising ever more questions: *How am I supposed to behave without a drink? Does the fact that I'm so bored make me a boring person? Why am I so socially awkward? Did we really only bond over booze?*

And let's pan out again for a moment here. In the last chapter, I mentioned how I believe our Sober Curiosity is partly a response to the rapid pace of change that's the hallmark of early twenty-first-century life. How there's a growing sense that this is no time to be getting *out of it*—since our future thriving as a species will require all of us to be very much *in it*: having our say, shaping policy, taking better care of the environment, and consciously creating more inclusive communities, for example. You could say that globalization, advances in technology, and the resulting social and political unrest have created a *collective edge*—and one that's got us all asking some pretty big questions. *What are our values as a society? What's been* my *role in creating the problems we're faced with? Does this make me a bad person? How can I become part of the solution instead?*

You need a pretty clear head to begin to answer uncomfortable questions like these. Not to mention be fully present and prepared to act with as much integrity as you can muster. In his TED Talk, Johann Hari says that "a core part of addiction . . . is about not being able to bear to be present in your life." Which also begs the question: To what extent does our *addiction to comfort* reflect our inability or unwillingness to bear confronting the answers to these questions, and more?

In *How Soon Is Now?: From Personal Initiation to Global Transformation* (2017), his book on what he sees as nothing short of impending social, political, and environmental Armageddon (it's actually a progressive and inspiring read!), philosopher Daniel Pinchbeck likens our current situation to a *mass initiation* into a new era of consciousness. "All traditional societies around the world—all premodern cultures—had some form of rites of passage, of initiations, which marked the transition from childhood to maturity," he writes, suggesting that the edge we're on, or perhaps even the *sheer cliff face* we find ourselves staring down, is actually an invitation to assume individual, grown-up responsibility for where we go from here.

Initiation ceremonies in many cultures often revolve around the initiate facing some kind of potentially mortal danger as a way of proving "I can take care of myself now." What does it say about what we know, deep down, to be the "potentially mortal danger" of consuming ethyl alcohol when *getting trashed* when we turn twenty-one has been adopted as our Western initiation of choice?

Could it be that getting Sober Curious—facing the perceived danger of being the odd one out, risking social suicide, and being prepared to brush up against our spikier emotional edges—is also an initiation of sorts?

The Sanskrit word *tapas* means "to heat," and when used in yoga it speaks to the practice of "standing in the fire for the sake of positive change," says Stephanie Snyder. Literally, this means holding a difficult pose, feeling the burn as muscles tighten and contract, maintaining the mental focus it takes to stay in the pose—and choosing to endure all this in the name of becoming stronger, more agile, *unfuckwithable.* Applied metaphorically to our Sober Curiosity, it means sitting in whatever WTF we happen to be experiencing as a result of not drinking, watching it pass, and choosing to focus on the positive parts of the experience. These positives might include learning who your real friends are; giving less and less of a fuck about trying to "fit in"; reframing your workaholism as getting-super-productive-with-my-actual-life-goals; only sleeping with people you really want to sleep with—among others that will be very personal to you!

In the meantime, I repeat: whatever WTF you happen to be experiencing as a result of not drinking, *it will pass*—with patience, perseverance, community to lean on, and plenty of self-love, that is. Snyder says that "somebody who is self-medicating any chance they get is basically saying, 'I have no tolerance for discomfort.' Which is ridiculous, because there is a lot of discomfort in life."

Are you comfortable with this idea yet?

The Future Is Emotionally Intelligent

I mentioned that my natural emotional resting state could be described as "mildly tense but hopeful for a positive outcome." This mild tension that I've identified, accepted, and even *embraced* as part of my personality since getting Sober Curious is one of the core feelings I used to drink in order to numb—because it was a part of myself I found uncomfortable to bear.

A serious, bookish child and a grade-A student in school, I went on to choose a career that revolves around meeting persistent, weekly deadlines. As a freelancer (living in one of the most expensive, competitive, and fast-paced cities in the world, no less), I also never know where the next job or paycheck is coming from. Enough to make anybody tense—even if the privileges of my background, my education, and the color of my skin lessen the daily load on my nervous system.

Taking the mild tension part into account, it makes sense that one of my most lingering FOMA triggers is reaching the end of a complex writing project, nailing a tight deadline, or receiving a check I know will cover next month's rent. In these instances, the illusion is that a drink will enhance the feeling of *giving-very-few-fucks*, which is basically the opposite of my daily outlook. That a drink will switch off any residual "stress" and thereby prolong the experience of ease that accompanies all of the scenarios listed.

This process is an example of my having developed *emotional intelligence* around one of the reasons I drank. And am *still* sometimes likely to crave a drink, even having learned from repeated, painful experience that the glass of wine that I'm hoping will ease my tension will eventually only exacerbate it. How can I use this information? Prioritize and engage in other ways of mitigating tension—such as doing yoga or getting a regular deep-tissue massage—before the FOMA hits.

Emotional intelligence is not something we're taught in school, and in fact we live in a society that tends not to have much time for feelings at all. They may be the messengers of our soul, "an essential part of our very accurate sensory system," as Guru-Meher Khalsa puts it, but feelings are also deeply unpredictable, utterly irrational, potentially overwhelming, and not exactly prone to keeping to any kind of schedule. In a world where time is money, addressing feelings can often be seen as more expense, and trouble, than it's worth.

But as we've already seen, feelings don't like to be ignored. Especially if they are delivering what they see as VERY IMPORTANT INFORMATION concerning our well-being. As this relates to your Sober Curiosity, drinking and using other substances can be both ways to numb out from persistent, unmet emotional needs *and* outlets for those emotional needs to be expressed. How many times have you gone out in the evening seeking an escape from stress, anxiety, or sadness and finished the night crying into your wine as the very same stress, anxiety, and sadness surges out and overflows its banks?

The problem with feeling all your feelings this way is that it's like that overflowing *release* never really happened. *It's like cheating on an exam.* The only way to truly hear, learn from, and integrate what our feelings are trying to tell us is to sit with them *sober*—meaning with a clear, rational head—and really listen. Not to mention the fact that, as Marc Lewis writes in *The Biology of Desire*, "Research tells us unambiguously that suppression [of desire, cravings, and *our emotions*] is the wrong way to go, as it accelerates ego fatigue. The best way to resist temptation [rather than try to suppress desire] is to reinterpret your emotional state. Instead of tying yourself to the mast to resist the Sirens' song, you must recognize the Sirens as harbingers of death and reframe their songs as background noise."

Developing our own emotional intelligence can be the first step toward guiding ourselves to make life choices that support our highest good. As GuruMeher Khalsa writes in *Sense of the Soul:* "By consciously listening to your feelings, even when they are unpleasant, you will discover what is wrong, what you need, and exactly what you can do about it . . . When you respond not from the emotion but instead act through your consciousness—even though what needs to be done is daunting—you take care of yourself and handle your life. This restores confidence and then trust in yourself."

How to go about this? Seeking connections with individuals you feel safe to be your whole self with is step number one. These are individuals whom you trust to mirror back to you that whatever you are feeling is OKAY. My work these days means I have found these individuals by attending healing circles and moon rituals and practicing many of the alternative methods for emotional well-being that I write about for The Numinous. You may find yours in a structured support group or through pursuing a hobby that you now have TONS of time for since you no longer spend half your weekend nursing a hangover. With a regular meditation practice, eventually you may also find that one of these supportive individuals is YOU.

And speaking of emotional healing, "talking it out" is great, and an essential way to *name and claim* our feelings. But it's also not the only way to process them. Stephanie Snyder points out that with yoga, for example, you might experience "deep and profound understandings that come spontaneously" through the practice. You might not always know the whole story behind these understandings; "it could be a few tears," she says, "or some little shivers you might feel during the practice that are the emotional release."

Any time you become present to an unwelcome emotion, it's also important to resist the urge to immediately back off from it. How can you become familiar with all your different feeling states unless you're prepared to linger a while and get to know them?

Okay, deep breath.

If all this feeling of feelings is sounding a little intimidating, then let's take a moment to remind ourselves that it's all in service to acknowledging, accepting, and ultimately resolving the root of the heavier emotions (which are not *bad* but which often *weigh us down*), and in doing so to allow JOY to bob back up to the surface.

All this talk of joy also makes me think of Marie Kondo's 2011 international smash hit *The Life-Changing Magic of Tidying Up: The Japanese Art of Decluttering and Organizing*. If you're one of about five people in the world who hasn't read it yet, Kondo leads her readers through a marathon tidying session, guiding them to remove from their lives all the material possessions that are clogging up the arteries of their existence—thereby freeing them to lead happier, lighter lives.

Her criterion for what to keep? Ask yourself the question, "What things will bring me *joy* if I keep them as part of my life?" Although it's a supremely practical book, feelings nonetheless "are the standard for decision making," with Kondo going so far as to claim, "If you act on your intuition [meaning, as I read it, *really pay attention to and act on the messages from your feelings*] you will be amazed at how things will begin to connect in your life. It is as if your life has been touched by magic."

Rather than seeing this part of your Sober Curious journey as being a painful deep dive into the murkier reaches of your psyche, you could choose to see it as the point where you get to "Kondo" your *inner world*. The emotional intelligence you will develop in doing this is also how you will empower yourself to experience a happier, lighter, and more confident existence. How you will begin to trust in *your own power over your own life*, proving to your Self that "I got this."

Our emotions may guide us to examine some uncomfortable truths, but ultimately, GuruMeher Khalsa says, "The goal is complete safety to be you and to handle all that this world brings you. We are all on a journey that includes learning to take better care of ourselves."

A journey that will make you unfuckwithable.

Alcohol and the Confidence Paradox

Another reason many of us, myself included, use alcohol is to feel more confident. More like "I got this" when faced with any number of situations where it feels like anything but. Walking into a room full of strangers you want to make a good impression on. Hitting the dance floor at a party. Especially one of those dance floors where people form a circle and stand around whooping and cheering for the person throwing shapes in the middle. Jeez. Moving a juicy date into the zone of physical intimacy. Speaking your truth when what you have to say may needle a past pain, go against the grain, or lead to a confrontation.

You may be familiar with some such situations and have used booze to enable you to participate. But alcohol doesn't *give us* confidence in situations like these; all it does is suppress or numb our fear. The *confidence* we crave is a feeling state that can be cultivated only within ourselves.

How do we do this? Some generally acknowledged ways to increase confidence include:

VISUALIZATION. Create an image in your mind of a more self-possessed version of you doing all the things you're afraid of, confidently. For example, every time I have a public-

speaking gig looming, I take some time during the days leading up to it to picture it going well in my mind's eye. You can amplify the effect by attempting to re-create the positive feelings associated with this.

POSITIVE AFFIRMATIONS. Repeat positive and uplifting statements about your life daily, ideally out loud. This can sound and feel very silly, but it is essentially a way to reprogram your negativity bias, so it's worth sticking with! Try saying them to your pet or to your reflection in the mirror to make it feel less awkward. Plus, pro tip: Phrase the affirmations as questions for maximum effect since our brains love to go on the hunt for answers. For example, "Why am I so great at public speaking?"

POWER POSES. Stand up tall, shoulders back, head up. Like Wonder Woman. Research by social psychologist Amy Cuddy has shown that adopting a confident stance can release testosterone, making you feel braver and more confident. "We convince by our presence, and to convince others we need to convince ourselves," she writes in *Presence: Bringing Your Boldest Self to Your Biggest Challenges* (2015).

MODELING BEHAVIOR. Choose a "confidence icon" (somebody whose confidence you admire) to channel when doing things that scare you. The accompanying affirmation might go, "Why am I as good at public speaking as Oprah Winfrey?"

HELPING OTHERS. Volunteering, mentoring, or practically assisting somebody in need automatically shifts your perspective outward from "self-pity" to "compassion," while also helping you feel thankful for what you have.

But the ultimate trick to building confidence? *Feel the fear and do the scary thing anyway*—and in doing so, with all due respect, prove the fears wrong. Which, by the way, is not the same as holding your breath, closing your eyes, saying a prayer, and diving headfirst off the *sheer cliff face*. After all, fear and anxiety (the emotions that stop us from diving headfirst into what we perceive as being dangerous situations) exist to keep us alert and to sharpen our senses. They exist to help us gather information about the nature of the danger we're facing, and then choose whatever action will protect us. My mildly tense outlook? It's also what keeps me alert and focused, helping me walk life's tightrope with grace and without having a meltdown every five minutes. These days, thanks to my Sober Curiosity, my meditation practice, and the resulting emotional intelligence I've developed, it's often little more than "background noise."

Action (the *doing it anyway* part of the above equation) is the key piece when it comes to resolving feelings of fear, and consequently proving to ourselves that "I *got* this"—thus cultivating the confidence, or self-belief, that I also got this the next time I am faced with said "dangerous" situation.

Which is where the "confidence paradox" I often talk about at Club SÖDA NYC events comes in. Feeling the fear and doing it anyway is basically what you're practicing each and every time you dive headfirst into another Sober First. If we've been led to believe, through both external conditioning and our own experience suppressing fear with alcohol, that drinking is what makes us confident, then getting Sober Curious and choosing to acknowledge our fear, asking it what it has to say, and acting on this accordingly is actually a fast track to growing our own *inner* self-assurance.

Which, again, is easier said than done. If you've been teaching your brain since, say, age fifteen (or whenever you first used booze to fend off fear) that the way to feel more confident is to drink, then this will be a pretty deeply ingrained neural pathway to reroute. To do it you'll have to repeatedly and determinedly go against the grain and make a different choice. You'll have to commit to sitting in the discomfort of this process in the name of positive change.

This was a process I had been practicing intuitively for several years by the time I found my way to AA. The reason I found it so hard to "admit" that I was an alcoholic? That I was "powerless over alcohol"? Because the statement is essentially the opposite of a positive affirmation. By choosing, one dry dance party and dinner date at a time, not to drink, my Sober Curiosity had already proved to me that I was FULL of power. That I was the mistress and the respectful student of all my differing feeling states. Also, that I was chattier, funnier, and a better dancer than I had ever *trusted* myself to be.

If waking up without a hangover feels good, then try waking up the morning after proving to yourself what a social, funny, and groovy human being you are without "needing" a drink. Just by being YOU. That's a punch-the-air *joyful* way to start your day, right there. Feels to me a lot like *acing an exam.*

Of course, it may not feel as if you pass every Sober First with flying colors. Sometimes you will feel exactly as awkward and unfunny and like as bad a dancer as you feared. You'll be bored by the small talk, will fail to articulate your truth or have it received in the way you hoped. On occasions like these, any fear, discomfort, and anxiety that arises is likely telling you that the next best action to take is to take yourself home, to try again with the difficult conversation another day, or to attempt to express your truth in another way—even if this makes you seem weird, like an outsider, or even kind of a bitch.

This is where we learn to quiet the *right temporoparietal junction*—the part of our brain that monitors what other people think of us—for ourselves.

The reason I feel more confident as Sober Curious me than I ever did as drunk me? Because I have also learned to *trust and appreciate* that whatever feelings I am feeling—and whatever words I am speaking and whatever actions I am choosing as a result—support my highest good. That they are ultimately in service to clearing the path, again and again, back to joy.

GETTING HIGH ON MY OWN SUPPLY

The tagline for my brand The Numinous (and the subtitle of my first book *Material Girl, Mystical World*) is *The Now Age Guide to a High-Vibe Life*. "Now Age" because my mission is to take *New Age* practices like astrology, yoga, meditation, and energy healing and show how relevant they are for *now*. On the basis of the impact they've had on my life (including helping me get Sober Curious), I've come to see these tools as essential components to navigating and balancing our sped-up, disconnected, robotic, and often highly medicated experience of life in the twenty-first century. "Now Age" also refers to the culture of instant gratification that's rocket fuel to our addictions, and "High-Vibe" to the fact that the aim of many of these practices is to raise our energetic vibration—that is, the frequency at which the atoms in our cells are vibrating and communicating with the particles of source energy that permeate and hold all physical matter together.

And why would you want to do that? Because the higher our overall *vibe*, the more optimistic, the more peaceful, the more accepting, and the more *joyful* we feel. That's right: we experience the changes in our vibration as differing *feeling states.*

David R. Hawkins studied energetic frequencies for thirty years and gave feeling states a numeric value in what he called his Map of Consciousness. On his "scale of consciousness" from 1 to 1,000, shame bottoms out with a frequency of just 20, closely followed by guilt, at 30. Moving up the scale, courage comes in at 200, love at 500, and joy swoops in at 540. Beyond 700, meanwhile, is where "enlightenment" lies.

Seeking higher vibes does not, by the way, detract from or disavow lower vibrational states (our experiences of shame, guilt, anger, and sadness, for example). Low vibes are as real and as valid as high vibes, since *all* our feelings contain valuable messages from our soul. The good news is that, because we are conscious beings, it is within our power to listen to and learn from the lower vibes in order to move toward the higher vibrational states. Meaning, it is in our power to make choices and take action to heal, or reconcile, whatever led to the low vibes taking hold.

As for *how* we might do that?

"Conscious" is the key word here. The more *consciously* we choose which thoughts, beliefs, and judgments—not to mention which pieces of information, which human beings, and which *toxic mood-altering substances*—we allow into the energy field that surrounds us, the more influence we have over our energetic vibration. Therefore, living a "high-vibe life" simply means seeking to raise our consciousness so we can be better informed about which next choices to make, and therefore which next actions to take, in service to our overall well-being.

Whew! Which possibly just went wayyyyy woo-woo for some of you. So, let's press "pause" on the vibes for a moment and consider what "consciousness raising" might mean in relation to what we've learned thus far about the impact of alcohol on *how we feel*—whether in the pages of this book or embedded in the unassailable truth of our own experiences.

I've spoken about the paradoxes that often muddle our relationships with booze. Well here's another one: *How come we drink to get "high" when alcohol is a known depressant?* Meaning, literally, something that pulls or pushes us into a "lower" state. Could it be that this is also referring to alcohol (despite the initial "high") lowering our level of consciousness? This is one way to explain the feelings of doubt, confusion, and regret (those "low-vibe" feelings again) that often accompany our hangovers.

But this doesn't fully explain my surprisingly joyful experience of not drinking—like the euphoric bliss I began to feel that was not attached to any "positive" event in particular, a feeling that had me dancing down the street when I first began to experiment with extended bouts of abstinence. See also the feelings of calm and "I got this" when faced with stressful situations. At first, it seemed as if consciously choosing not to drink had resulted in my feeling more naturally "high" all the time. As if these higher vibrational feeling states were actually my default settings, the place I was destined to return to once a depressant such as alcohol was removed. This latter theory fits with Hawkins's view, who writes: "The actual effect of drugs is merely to suppress the lower energy fields, thereby allowing the user to experience exclusively the higher ones . . . The suppression of the lower notes does not create the higher ones; it merely reveals their presence."

The English word "high" has its roots in ancient languages like Old Saxon, Norse, and Danish, where it was used in reference to all things large, momentous, or otherwise more *awesome* (in the biblical sense) than the norm. According to the Online Etymology Dictionary, it was first used to describe feeling "euphoric or exhilarated from alcohol" around 1620 and to describe the euphoric effects of other drugs in the 1930s. Which charts a well-trodden history of humans using alcohol to "access" the desired higher vibrational states. To seek to restore, perhaps, the "wholeness and peace of mind" that, as we have seen, Annie Grace asserts "we knew our entire lives before we drank a drop."

But one *great* thing about life in the Now Age—about getting to experience the dawning of the age of *information*—is that human consciousness is being raised exponentially. Thanks to the internet, we have the opportunity to be better informed and to become more aware than ever—all the better to decide for *ourselves* which next thoughts, beliefs, judgments, pieces of information, humans, and substances we invite into our energy fields. To an extent. Sometimes life circumstances out of our control will limit the degree to which we experience the privilege of choice.

But if the central tenet of many of the most progressive empowerment initiatives is *education*, then the more we can do to educate ourselves—beginning with getting *curious*—about the options available to us, and to those in our wider communities, the better. The more assistance and positive role models we can then gather to affirm that we are, indeed, supported in this endeavor, the more we begin to trust in *our own power over our own lives.*

Here's an example of how this might impact your boozestory, the narrative you may have held to be true if you're a regular ol' moderate-to-heavy social drinker like I was. At the most recent Club SÖDA NYC event, a guest raised her hand during the Q&A to ask: "What are your views on moderation? Is it possible to cut back and just drink less?" My response went something like this: "The question of moderation is relevant only if you still believe that your happiness is connected to alcohol—if you still trust that alcohol is how you experience relaxation, connection, inspiration, and joy. If you're at this event, then it's likely that life has shown you otherwise. Getting Sober Curious is about learning to trust the truth of your experience and your body over all the external messaging about booze. What do you *truly* need alcohol for, even in moderate quantities, anyway?"

Our events, I went on to explain, are designed to remind people that it's possible to generate all the highs we could ever need for ourselves. To rocket our own vibration into the stratosphere, without the side of existential misery.

Of course, the next question is generally along the lines of, *Well what else can I do to get high?* Which is where we're headed next.

Sing, Dance, Tell Stories, Savor the Sweetness of Silence

In the last chapter, I told you how my best friend's response when I told her I was thinking about quitting drinking was along the lines of "But why?! You only drink to have fun!"

She was possibly remembering the time on a rooftop bar in Ibiza when we were laughing, dancing, and having so much "fun" with our friends that a couple who'd just gotten engaged sent *us* a bottle of champagne. How we went on to serenade them at the tops of our lungs, the bubbles needling our nostrils, hearts beating out a rhythm of pure love. We sang and danced *a lot* on that vacation, if I remember rightly, fueled by shots of medicinal-tasting local Hierbas liqueur and fiery mojitos on the beach. Heading to the airport in the glow of another heart-wrecking sunset, I remember waves of gratitude breaking in my chest and throat as tears streamed down my face. What a *gift* it was to be alive.

A moment of *euphoric recall* that also reminds me of a story cultural anthropologist Angeles Arrien tells in the foreword to Gabrielle Roth's book *Maps to Ecstasy: A Healing Journey for the Untamed Spirit* (rev. ed. 1998): "In many shamanic societies, if you came to a shaman or medicine person complaining of being disheartened, dispirited, or depressed, they would ask you one of four questions: When did you stop dancing? When did you stop singing? When did you stop being enchanted by stories? When did you stop finding comfort in the sweet territory of silence?" Arrien calls these the "four universal healing salves."

I believe that dancing, singing, stories, and silence (or meditation) are some of the languages of the soul, as well as Arrien's healing salves. Her story implies that sadness, or feeling downtrodden by life, is the result of our having stopped listening to these languages. Which often happens at whatever age we begin to consider what it means to be a "grown-up." Singing, dancing, storytelling, and afternoon quiet or "nap" time are all common nursery school activities, after all, which often get nixed once the serious business of getting trained to be a good little worker kicks in. Could it be that reinstating these activities, rediscovering these salves, and learning to speak these languages again, are steps in restoring the *wholeness and peace of mind* that are our birthright?

It's also telling that dancing, singing, and storytelling (perhaps in the form of some juicy Saturday brunch-time gossip) are some of the activities we've learned, as grown-ups, to outsource to booze. Are some of the first things we want to do once the second or third drink has kicked in. As for the *sweet territory of silence*, drinking alone is often considered a sure sign that a person's Pinot habit has veered into risky drinking territory. But even sitting here now, when I've been Sober Curious for almost a decade, the memory of a glass of red taken solo in the perfect stillness of a rainy afternoon, the air so thick it feels like a cashmere blanket, vibrates with an almost religious quality for me.

But the Truth-with-a-capital-T is that we don't need *alcohol* to experience any of these things. Even if it's equally true that when we've been teaching our brains otherwise since whatever age we began acting like grown-ups, our souls may have some work to do when it comes to persuading our minds and bodies otherwise.

If living a fully rounded, whole, and unexpectedly blissful Sober Curious life is about *consciously choosing* to remove alcohol from the equation, then it is equally about making proactive choices about what to invite back *in* to your life. Remember what Rosie Boycott wrote about "dry drunk" syndrome? "Working on the premise that life without booze must be better, people stop drinking and wait for their worlds to improve, as though they are owed a favor for having made the supreme sacrifice of cutting out the sauce."

Life without booze WILL be better. Trust me. But if you're ready to get *high* on your own supply, you can't just sit there. It's time to ask yourself the question: *What truly makes my vibes come ALIVE?*

Highs Worth Having Are Worth Working For

> ***I have never in my life envied a human being who led an easy life. I have envied a great many people who led difficult lives and led them well.***
>
> —THEODORE ROOSEVELT

Some of the higher vibrational feeling states I used alcohol to attempt to access were relaxation, amusement, connection, pleasure, and transcendence. The good news is that I experience all these "highs," and more, as a Sober Curious person—even if the ways in which I experience them often look a little different and are rarely as fast-acting as downing a Tequila Slammer. As I mentioned, the term "Now Age" also describes a culture of instant gratification that's tricked us into believing we can have access to anything and everything we need, including feeling states, at the touch of a button (or the popping of a pill or the downing of a shot). Not true.

The beginning of the above quotation by Teddy Roosevelt goes like this: *"Nothing in the world is worth having or worth doing unless it means effort, pain, difficulty,"* and I'm not about to suggest that generating your own bone fide high vibes is always going to be *painful* or *difficult.* But it may well require a little more *effort*—including the effort of undoing the conditioning about which people, parties, and other pastimes make you feel good. Which in part, as discussed in chapter 2, will mean undoing the social conditioning as to what constitutes an exciting, fulfilling, and glamorous social life.

Not to mention that natural highs are way more subtle, ephemeral, and harder to quantify than chemical highs. It's not always as simple as "perform *X* activity and you will feel *Y.*" If you've been used to dowsing the innocent green shoots of your natural highs with rocket fuel, it may also be harder to feel them at all at first—the same way you're less likely to appreciate the delicate flavors of a plant-based feast if your taste buds are used to getting jacked on a diet of pizza and Mickey D's.

What I'm saying is, as you peruse the following pick 'n' mix of some of *my* favorite Sober Curious highs, try to keep an open mind and heart. Notice any judgments that come up, and consider what shameful past experiences or prejudices are perhaps being triggered. Above all, be willing to imagine what a brave, new, booze-free world, laced with laughter and uniquely catered to YOU, might look like. In the words of Recovery 2.0 founder Tommy Rosen: "When [a person has] organized their life around a substance in such a way that the only time they have fun is when they're engaged with that substance . . . when the substance goes away, at first one might think, 'Oh my god. The party's over.' But the party's not over at all. The party's just beginning, but in a different way than what you're used to."

When I Want to Relax

GETTING OFFLINE

Last Christmas, I accidentally booked an Airbnb with no WiFi on a vacation to Joshua Tree National Park—and relaxed more in twenty-four hours than I had in the previous five years. I experienced a level of calm that felt like something had unclenched on a cellular level. That felt *blissful.* Key to this phenomenon was my not being able to access my Instagram account (which I use predominantly for work), which has led me to consciously choosing to delete the IG app from my phone weeknights and all weekend and thereby experience said blissful unclenching on a regular basis.

For you it may be Facebook or YouTube, each new post or video tempting you to squander another small piece of your attention, or consciousness. Each being another small piece of information to process, creating an emotional reaction in your body that your mind will then be tasked with responding to. Exhausting! Ditto email, which I already put strict boundaries around to enable me to write. Why? Writing requires single-minded focus—which is the exact opposite of constantly checking, and being triggered by, your inbox or social media feeds.

CRAFTING AND CROSSWORD PUZZLES

These are two ways to distract my workaholic monkey mind—the part of me that wants to do, do, do which has helped me put so many of my dreams into action, and which also needs to be tricked into doing something else sometimes in the name of my sanity. One Sober Curious friend tells me that crafting (in her case this usually means making visually satirical signs to take on protest marches) satisfies the exact same part of her brain as having a drink. I think what she means is that creating something with her hands is like giving the monkey mind a toy to play with (a toy that may previously have come in a frosted glass with a lime-salted rim) while she (her higher Self) gets on with the serious business of *chilling the fuck out*. If you're a word person like me, crosswords can be toys, too. You may also find that coloring books, collaging, and cooking work a treat.

CRYING

I'm a big crier. HUGE. I cry when I'm happy, when I'm sad, when I'm outraged, when I speak up about anything I care about, when I sing, when I dance . . . but I realized recently that I started crying regularly only *after* I got Sober Curious. Also, what a relief it is—literally how *relaxing*—to let the tears flow. I used to suffer a lot from sore throats, particularly when I was anxious or upset, and I believe this was from subconsciously contracting the muscles in my pharynx to stop the tears from leaking out. How *shameful* to be seen as a big old cry baby. And how different would the world be if we were taught to welcome our tears as a *release?*—of tension, of pain, of anger, or of anything else that prevents us from accessing our natural *resting* states of wholeness and peace of mind. When you think about the sheer weight of water, doesn't it make sense that offloading a few bucketloads of tears will automatically allow your vibe to rise?

An extreme example of the healing power of tears, and of it being "okay" to cry, is shown in a 2017 documentary called *The Work*, which follows a group of men facilitating emotional healing work with inmates at Folsom State Prison in California. Their subjects are some of the most violent criminals in the country, who, one by one, are finally given space to share the extent of the pain that led them to pursue the paths that landed them in jail. Absent fathers. Hopeless poverty. And tale after tale of learning that being "a man" meant being hard and not showing emotion. When these men allow their tears to break them open, the release is like a thunderstorm. Violent, torrential, cleansing. And the aftermath is a sense of all-pervasive *calm*. In the twenty years that the program has been running, not one of the inmates who has participated and been released from prison has reoffended.

When I Want to Have Fun

SINGING

I "stopped singing" around age thirteen after I discovered at a friend's karaoke birthday party that my singing voice was not quite as Madonna-esque as I'd imagined. The handful of occasions I'd gotten back into the karaoke booth, drunk, naturally, had only dredged up the earlier feelings of shame—a spell that was broken on my first Sober Curious bachelorette.

When I saw a karaoke bar on the itinerary, I told myself it would be perfectly cool to take myself home at that point (NB, it totally would have been—reminder that this option is always available to you if and when you find yourself flailing halfway through a torturous Sober First). But when the time came I was already having so much fun I decided to dive in—and they could barely get me off the mic!—as I, fittingly, belted out a word-perfect rendition of Britney Spears's "Toxic" (who knew?). With no numbing substance in the mix, I could *feel the vibrations rising in my body*—beginning in my heart and lungs, spreading across my face, and engulfing my head in happy tingles.

This is a phenomenon I have also experienced singing hymns on the rare occasions I've been to church, throwing my voice into the crowd at concerts, singing along full volume to the radio while driving, and chanting *Om* (like I mean it) in yoga class. There's something about raising your voice in song, alone or with others, that feels like joining in with the rhythm and rhyme of the universe. A feeling you can *really feel* when you're sober.

DANCING

Something else I never thought I would be able to do stone-cold Sober Curious was dance my ass off. Even if, after my mom took me to see the movie *Grease* when I was three years old, I spent six months performing the routines and declaring to anybody who would listen: "Ruby is truly a dancer!" But since my teens, this is something else I had outsourced to alcohol and other drugs—

until my breakthrough moment at the New Moon rave in Croatia I wrote about in chapter 1. What had shifted?

As we now know, a big part of getting Sober Curious is getting more accepting of our whole self. Including the wild, sexy, attention-seeking self that gets activated when music makes us want to *move.* In a society where wild, sexy, attention-seeking is simultaneously censured and celebrated (meaning it's all over our media but often seen as just plain tragic in the rest of us), such as in the US or the UK, this self is one of the most-often shamed and therefore most suppressed, meaning we often have to disable our right temporo-parietal junction with booze before it feels safe to let loose. In Croatia, the Sober Firsts I'd notched up to that date meant I'd gotten okay enough with my whole self, including my wild, sexy, attention-seeking, dancing self, to *just go for it.* And even better, my going for it seemed to give everybody else permission to go for it, too.

Before you try this one, it may help to get some practice. For example, any time you get a surge of Sober Curious joy while you happen to have some killer tunes playing, try dancing wildly on your own and *just going with it.* Close your eyes if it helps, and really pay attention to how good it *feels* to move your body to the music whichever way it wants. This could be enough for you! But eventually you may be ready to unleash your wild, sexy, attention-seeking self in a proper party scenario and disable the booty-shaking shame once and for all.

LEARNING

The fun factor tends to get stripped out of learning at school as soon as it becomes about gold stars, grades, tests, and competition. Sure, competing can also feel like fun . . . when you're winning. But the flip side, the losing side, can also unleash waves of shame. Thanks for making this dynamic the basis for society as we know it, capitalism. Learning for learning's sake, however, meaning simply following your curiosity about a subject that activates the *vibration of fascination* in you, can feel a lot like falling in love.

Once the clarity of my initial Sober Curious awakening found me questioning my dream job at the *Sunday Times* (*"Is this all there is?"*), the next thing I asked myself was, *"What's the one subject I could research and think and talk about all day, and never get bored of?"* The answer was *"Astrology."* Which is what led to my starting The Numinous, experiencing my spiritual awakening, writing about it in my first book, getting properly Sober Curious, and, well, here we are. Since "having fun" could also be termed "not being bored," inquiring into what *you* could research and think and talk about all day and not get bored could be a helpful starting point for you if it feels like Sober Curious monotony has set in.

Getting offline is a great first step in rediscovering what this might be, since this allows you to see what path your curiosity naturally wants to lead you down, versus having it pulled this way and that by endless new memes, headlines, and cat videos. And I say "rediscovering" because there may also be some clues in whatever it was you loved to do as a kid, or before life became mainly about homework, fitting in, and making sure the bills got paid. For example, during our WiFi break at Joshua Tree, the Pisces remembered how important DJ-ing had been to him. Since he had gotten rid of his record collection when he got a "proper job" and began to focus on climbing the corporate ladder, he has invested in some digital turntables. And considering how much I love to dance around our apartment, the fact that he now spends every Sunday making mixes is a major win-win for the high vibes!

ADVENTURE

It's a scientific fact that taking a risk increases production of the feel-good chemical dopamine in the brain, with a chaser of energizing, pain-killing adrenaline. As does drinking alcohol and taking other drugs. The good news is that replacing the latter with the former does not have to look like endless bungee jumps or shark-tank safaris. Going on "an adventure" can be as simple as taking a walk around your city without your phone. Volunteering at a homeless shelter. Challenging yourself to get to know someone without asking them what they do. Any and all of your Sober Firsts. Considering the extent of our modern comfort epidemic, it may be surprisingly easy to find experiences that take you out of your comfort zone and into the adventurous realm of risk-taking and ramped-up vibes.

Another plus is that doing new things slows down our perception of time passing, making life appear fuller and more multifaceted. This is because when we're engaged in a novel activity, or adventure, our brain is fully occupied with laying down new memories, taking time to pay attention to each and every detail rather than skipping ahead to what it knows is coming next. Only really possible when we're Sober Curious, too, since the part of our brain that creates new memories is disabled when we drink—

and hands up, those of you who have used booze to *kill time* when you're not having fun.

When I Want to Connect

COMMUNITY

Being in community means not only being around other people, but being around other people who share your interests, challenges, goals, and beliefs. You may find community in starting your own Sober Curious support group. Or at AA, if your boozestory is leading you that way. Also, in sports teams and book clubs and writing groups and activism. What you decide you want to learn about could help you find your community too. Beginning my study of astrology gave me an excuse to connect with other people I felt a mystical affinity with. Other people with whom I was able to connect socially over more than cocktails and gossip, and whose own high vibes helped to raise mine.

Because another thing about vibrations? In quantum physics, a higher vibrational structure (including a higher vibrational human being?) is said to raise the vibration of any lower vibrational structure it comes into contact with. So, y'know, feel free to curate your community carefully. Also, no more *subconsciously lowering your vibe with alcohol* to stay at a level that makes others in your orbit more comfortable. Rise up, and those who are ready will rise with you.

TRUTH-TELLING

Truth-telling takes a lot of effort, considering all the ways we routinely shave the edges off our personal truths in the name of "keeping the peace," "fitting in," and "not disrupting comfortable-if-dysfunctional family and social dynamics." Such things are not the same as outright lying (obviously low-vibe territory), but I've awoken to the untruths of omission only since becoming fully conscious of all the things that have gone unsaid in my life, the truths that have remained unspoken. Which could be as simple as not telling my husband what's *really* been stressing me out (for instance, our not having enough sex—although voicing this did *not* feel simple at the time) or as complex as explaining to my mom the extent of my ordeal with the Capricorn (which I only had the courage to share twenty years after I ended the relationship)—the kind of personal truth bombs we might have gotten used to detonating while drunk, if at all, leading to all kinds of confusion, and which may have been weighing on our vibe for years.

SERVICE

What's the glue that holds any community together? Our willingness to help one another and a commitment to contributing to the greater good. A desire to be "of service" is both a commonly reported side effect of quitting booze *and* something that research has shown helps people stay sober longer—the theory being that when we're engaged in any "addictive" behaviors, our world pretty much shrinks to the size of whatever it will take to get our next fix. Resulting in varying degrees of self-absorption, depending on the level of the addiction and what it is we've become addicted to—social media and video games being two very common examples of things that routinely pull us away from contributing to society on a deeper level.

Pulling my head out of the ditch of my regular ol' social drinking habit resulted in my considering how my work might actually serve others (versus a lot of my magazine career being focused on running with a cool crowd and getting external validation about my own worth). Then there was the feel-good factor of mentoring young people in the Numinous community, volunteering, and using my platform to speak on issues I think are important. Activities which have been shown to be as beneficial to my health as stopping smoking.

Being "of service" can also be as simple as really being there, meaning being *fully present*, for a family member or friend in need. It could also mean offering to cook a meal for a colleague you know is overwhelmed or creating a series of inspiring Instagram memes. Remember, starting "local" is the first step toward creating bigger shifts on a global scale.

When I Want to Feel Pleasure

FOUR BODIES WELLNESS

Four bodies wellness means paying attention to our health on four levels: physical, mental, emotional, and spiritual. For me, a sense of overall well-being kicked in fully only after I began to address all "four bodies" of my health—when I began to prioritize daily physical exercise as a way to wake up my *chi* (life force) and connect my body to my spirit, meditation to befriend my monkey mind, and got on board with the idea that "toxins" could be thoughts, feelings, and situations as much as substances.

For example, the gut issues I had suffered during my entire adult life went away fully only after I stopped drinking—despite my doing all the "right" things physically (eating all the fiber, quitting all the dairy, taking all the probiotics, etc.). Why? As well as alcohol itself doing damage to my gut, I believe it's only then was I able to feel, integrate, and fully "digest" the emotional and spiritual fallout of my teenage years. As for why this part comes under "pleasure"? Same reason I call booze-free sleep *orgasmic sleep.* Feeling "well" in all four bodies feels warm, tingly, satisfied, complete.

DELICIOUSNESS

I once stalled when the Pisces wanted to leave a party so I could have just one more glass of rosé. "It's just so delichious," I slurred, having already had way more than my fill. It's intriguing to me, therefore, that since I got fully Sober Curious, wine has begun to taste like poison. Still a highly seductive poison, but kind of sour and stinky like juice that's gone off. The kind of taste that actually, now that my taste buds have gotten their full faculties back, gets in the way of my enjoying dinner. (Which goes out to anybody who may be lamenting the loss of "a nice glass of wine" with food.)

At first, removing rosé (and hoppy IPAs, floral Sav Blancs, and herbaceous gin martinis) left behind a deliciousness gap that manifested as some serious sugar cravings. Something of a *pleasure deficit.* And part of finding new ways to experience deliciousness regularly has been reclaiming the pleasure of eating whatever the fuck I like, whenever the fuck I want. This is definitely not the same as "eating my feelings." I urge you, too, to commit to only *feeling* those. But, as noted, there have been periods in my life when I (as goes for the majority of women I know) had a lot of rules around what foods were "allowed." Post–Sober Curiosity (not least thanks to the whole "accepting my whole self fully and unconditionally" piece), these rules no longer apply. It began as a kind of "well I'm not drinking, so I *deserve* dessert/the bread basket/a full-fat Coke." Luckily (given sugar's highly addictive nature), the sugar cravings have since subsided, but the joy of deliciousness is something to indulge in every day. And farther along this path, I no longer feel I have to "earn" it, either.

SEX

Will I have more sex? Less sex? Will the sex be better? Given that getting Sober Curious means becoming fully present in your body—all the better to feel your feelings, identify your needs, and therefore home in on which are the next best actions to take—it makes sense that Sober Curious sex is definitely better sex. Also, learning to accept and express the wild, sexy, attention-seeking, dancing part of me means I've gotten better at listening to her needs and asking for them to be met.

Speaking of pleasure and *shame*, however, voicing my sexual needs, sober, was shockingly hard to do at first. It felt like being asked to stand up in front of my seventh-grade classroom and explain the purpose of the female orgasm. In a society with such complex and conflicting attitudes about and representations of female sexual pleasure, I'd gotten used to just *hoping* it would happen how and when I wanted. When your lover is a Pisces, there's a degree of intuition at work, thank Goddess. But hints and meaningful signs can get a girl only so far. Since I was outsourcing a large chunk of my pleasure quota to booze, I could also happily let it slide. Not so post–Sober Curiosity. Making sure sex happens how and when I want has become the most pleasurable part of my four bodies wellness regime.

When I Want to Transcend

MUSIC

In a logical, linear, left-brain world, we are *starrrrrrrved* of right-brain experiences that transport us into the mystical. We are *crrrraaaaaving* opportunities to escape the daily grind and just . . . drift. Just like alcohol, music is one widely accepted and socially endorsed way to go there—one reason music and booze have gone hand in hand since the dawn of pagan festivals. But because of its effect on the amygdala, the part of the brain that regulates emotion, immersing yourself in music alone can also be a way to feel feelings without the need for words. Making it truly the language of the soul.

Another reason I think the Pisces has rediscovered DJ-ing as part of his Sober Curiosity is that the sonic journey satisfies the part of him (of us, since he's doing a lot of the spinning in our apartment) that used to use booze to transcend the earthly realms. His picking up this hobby again has also led to our rediscovering our love of nightclubs—yes, sober! In our forties! I thought that quitting drinking (not to mention my age—oh societal conditioning!) meant my clubbing days were over, the experience having become so deeply entwined with drinking in my twenties and thirties. But reentering the scene booze-free, I have felt the same *magic* I experienced being completely immersed in music as a teen.

How to bring music back into your life ("back" because who *wasn't* obsessed with music in their teens)? Make it a priority, a part of *your* four bodies wellness regime. Download the Shazam app to help identify tracks that make your body sing while you're out and about. Subscribe to Spotify and use the "radio" function to help you find new artists to follow. And don't assume you'll feel awkward without a beer in your hand at a concert or club—in fact, make it *all about the music* and assume you'll be transported to somewhere beautiful.

HEALING (AND BREATHWORK IN PARTICULAR)

It was NYC yoga teacher Amanda Capobianco who first coined the hashtag #HealingIsTheNewNightlife after moving to Williamsburg in Brooklyn, where she expected to spend late nights listening to bands in grimy dive bars and instead found herself socializing at sound baths, moon circles, and breathwork sessions. The latter being an active emotional healing technique that's set to music and often results in a MAJOR energetic release accompanied by what I have come to term the "multiple cry-gasm."

I first tried breathwork in a teepee in a Brooklyn park, and it was the "highest" I had ever gotten when sober. Afterward, I felt like my head might float off down the East River, leaving behind a body that could happily subsist forevermore on hugs alone. But *during* the session, well, that's when things got downright trippy. I'm talking full-on, full-color visuals of scenes from throughout my life, and a physical awareness of the energy in me, around me, and that connects all living things. I've had similar experiences with Kundalini yoga, a style of yoga that focuses mainly on manipulating the breath in order to move "stuck" energy; with hypnosis; and during shamanic healing sessions that have revealed to me in no uncertain terms the multifaceted nature of human consciousness. No substances required!

MEDITATION

Yes, again. I'm suggesting *again* that meditation may well be a valuable addition to your successful (as in *joyful*) Sober Curiosity. On one hand, there's Tommy Rosen's theory that "the opposite of addiction is meditation"—since this is a practice that demands you simply sit and be present with everything that you are in this moment. On the other, this *everything* includes the numinous, mystical, transcendent part of you—the part of us all that, in an action-oriented, progress-obsessed, thinking and doing, left-brain world, is so easily ignored, suppressed, and otherwise confined to the margins of existence. The *feeling and being part* that alcohol enables, as it shuts down the thinking and processing part. So, meditate because it will help you become more and more accepting of who and all that you are, in the here and now. And also meditate because it's a free, VIP ticket to the place where it is completely acceptable to just BE.

• • •

So, there you have some of my favorite ways to raise my vibe sky high—some of which require more conscious effort to incorporate into my life, and some of which have begun to come to me as naturally as breathing. The key to discovering yours? Stay curious and make no assumptions about how a given activity or situation is going to make you feel. Commit to trying new things, and focus on the positive parts of any experience. And if the low vibes persist, sit with them, listen, and take your time to process what they have to say—for without the lows, it's true, there would be no highs, after all.

Wanna Get High?

Are you after more inspiration? Here's how my Sober Curious crew told me they like to get high:

"I love putting my phone away and walking outside, fully present and grateful for all that I am and all that is. Also, sending mental green light love beams to everyone I see along the way."

"Immersive art experiences."

"Petting all. The. Dogs. For as long as feels good."

"Adventure! Talking to a really cool, slightly intimidating group of strangers, dancing on the subway, going on an 'adventure walk' with no destination in mind . . . anything that gives me a little adrenaline rush and boosts my confidence."

"Exercise—especially yoga and spin. Yoga because it puts me in touch with my spirit and soul, and with feeling my Self in my body, and spin because it makes me feel sexy, alive, and awakened to my own power."

"Cacao smoothies."

"Learning about quantum physics (turns out, I don't need weed to go on a complete trip about the nature of matter, perception, and reality!), swimming and floating in the ocean, and, honestly, talking with loved ones about their hopes and dreams."

"Waking up on a Saturday or Sunday morning knowing someone out there is hungover and it isn't me!"

THE POWER OF POSITIVE DRINKING?

So why Sober Curious and not just Sober *Sober*?

It's a question you may well be asking yourself, having absorbed what I've shared thus far: how the impulse to drink is often a response to the marketing of alcohol as a cure-all for the problem of *being human*; how easy it is to become addicted; and how the pleasure of social drinking is often a function of this addiction. A scratching of the itch. Maybe you've been discovering for yourself how much fun it can be to socialize without alcohol, and how much more connected you feel to your loved ones when you choose to meet them with total clarity. Perhaps you've even begun to investigate what you might have been using booze to avoid and now feel committed to confronting this head-on.

If you're somebody who's never enjoyed or seen the point of drinking, then I hope you're already feeling more confident about your choice to stay sober and less likely to make excuses for why you don't drink or be made to feel like a loser or an outsider when everybody else is drinking. If you're a person in recovery, then this question—why not just Sober *Sober*, people?!—may even be why you picked up this book. Not because Sober Curiosity will ever be an option for you, but perhaps because you're curious about how the other half drinks. Or you're looking for additions to your sobriety tool kit, or new ways to frame *your* experience of addiction.

One thing I know for sure is that if you've been experimenting with extended periods of abstinence, then by now, you most definitely will have experienced the orgasmic sleep, the focus, the presence, and the deeper connections that await on the other side of alcohol. You will, I hope, have been finding plenty of other ways to get high on your own supply. So, considering what Holly Whitaker reminded us about booze—that what we're essentially dealing with is "a toxic substance that, by the way, causes weight gain and breaks your capillaries. And makes you do stupid things. That is actually a depressant, and also breeds anxiety. And makes for really bad, shitty sex"—then why, WHY, am I now going to include a chapter on *the power of positive drinking*?

I've been asking myself this question, too. Writing a book is kind of like having a really long, super in-depth conversation with yourself about whatever you happen to be writing about—in this case, obviously, the nature of Sober Curiosity. And this far into the process (the living it and the writing about it), I'm doing a really, *really* good job of convincing myself that there is no place for alcohol in my life. That all drinking is ultimately a downer and that being "Ruby, Sober Curious" may well mean total, lifelong abstinence for me.

And yet.

Life also just isn't that black and white (for me), and alcohol isn't exactly going anywhere. I have good reason to believe—based on my own boozestory, the nature of my curiosity, and not to mention the fact that (according to the National Institute on Alcohol Abuse and Alcoholism) 90 percent of recovering alcoholics "relapse"—that there are likely to be a few more "reminders" in my own kind-of-just-a-little-bit-addicted-to-booze future and more opportunities for me to study, up close and in graphic detail, the nature of the beast.

So, again, *why Sober Curious and not just Sober Sober?*

For one, I find it helpful to remind myself that alcohol is not, actually, the bad guy. The bad guys are the reasons we choose booze and us using it in ways that can become problematic. Not to mention a society that values displays of extroverted confidence over periods of introverted self-reflection. A society that cattle-prods us into one-size-fits-all boxes from before we can talk. That fosters puritanical attitudes about pleasure and a mistrust of transcendent experiences. And that teaches us to shy away from our more complicated feelings. If alcohol is essentially just a substance (albeit a highly addictive toxin with a phalanx of painful side effects), could it also be possible to learn ways to coexist alongside booze without glorifying or demonizing it?

For two, there's the charge that tends to build up around anything that we prohibit ourselves from experiencing directly. Particularly something that, if we recall correctly, has been fairly effective in the past at providing us with pleasure while removing our pain. Remember that thing called ego fatigue? What Marc Lewis describes as "a loss of top-down cognitive control, augmented rather than diminished by attempts to suppress impulses"? In the case of somebody committed to changing deeply ingrained habitual drinking patterns, could it be that *consciously choosing* to drink on occasion—with the aim of unmasking the monster once and for all—is actually a way to accelerate past this pothole in the road?

In this case, a "positive" drinking experience may not always look like it does in the pictures (you know, the ones on Facebook of you having the time of your life with the new best friends you just met doing shots). It may feel, in fact, more like the worst decision you made all month. But can it become a positive if it goes some way toward re-minding yourself (as in *re*-programming your *mind*) to think differently about your *future* relationship with booze? As I discussed in the first chapter, each of my controlled relapses/reminders has helped inform longer and longer periods of easy-to-maintain abstinence.

What positive drinking is *not* (and I'll say it again) is making rules about when and why and how much you're "allowed" to drink (including deciding that "high quality" alcohol is somehow okay and won't make you feel as shitty as the cheap stuff). Reserving booze for certain occasions ("only on weekends/with fancy meals/with certain groups of friends") keeps it snugly positioned on the pedestal marked "pleasure." It becomes something you will then be more inclined to reach for any time you feel a pleasure deficit in your life (sadly, this means frequently for many of us), regardless of what you know by now about the less-than-pleasurable consequences. If you persist in regarding alcohol as desirable, then the desire function of your brain will eventually and inevitably push you to drink as much as you always did.

"Special occasion" drinking also begs another question: *Is it the occasion or the drinks that are special?* I tried this one for a while and discovered that using booze to "wet" landmark events (my fortieth birthday, for instance, or landing my first book deal) actually made them feel *less* special—kind of flat and two-dimensional. Not to mention that the memories are so sketchy. I can picture the scenes in my head, but I can't recall how these moments *felt* and therefore everything they *meant* to me, can't recall their impact on the ongoing story of my life. My euphoric recall of the boozy Ibiza vacation in the previous chapter? What's imprinted the most vividly in me is not the cocktails, but the feeling of being connected to my friends, the importance of taking a respite from the grind, and the heart-expanding wonder of spending time in nature. And *that's* what I want more of in the future. None of which requires alcohol.

But for now, in honor of the *curiosity* part of getting Sober Curious and with respect for your right to choose what's best for *you*, let's consider the situations in which this *might* mean choosing booze. As a ritual, perhaps, as a reminder, or even as one step closer to choosing full-blown, lifelong abstinence. And if you're already like, "Nah, thanks anyway but I'm done," then feel free to skip right on ahead to the next chapter!

Drinking, with Due Respect

Depending on how up to speed you are with all things Now Age, you may or may not be aware that the current drug of choice among spiritual seekers is a brew called ayahuasca, made from a medicinal South American vine. Administered by a shaman as a foul-tasting tea, "aya" is taken ceremonially to bring about deep healing of the psyche and the self. It is said to be particularly effective in treating PTSD and depression and in helping people realize the roots of their addictions—therefore equipping them with some tools to go about resolving them. A highly potent psychoactive drug, it also brings on deep hallucinations, can cause extreme emotional reactions, and often leads to vomiting and other means of purging—the latter being a part of the overall aya experience that's known as "getting well."

This is all secondhand information by the way, as I'm petrified of trying it myself. I prefer subtler means of higher Self-inquiry, such as meditation, breathwork, and hypnosis. But I mention it here because, as I just said, it is a highly potent psychoactive drug that can cause hallucinations, extreme emotional reactions, and vomiting—*just like alcohol*, in the right quantities. And yet, you don't find too many people drinking aya just because it's Friday night and everybody else is (although you do find some). For most, embarking on an ayahuasca trip often involves months of planning, some serious psychological and physiological prep, and a journey to the midst of the Peruvian jungle. It's a once-in-an-awakening kind of a thing.

Besides the physical effects I've highlighted, however, aya and alcohol are two completely different substances, and I am most definitely *not* suggesting that alcohol can be used to bring about deep psychological healing. As we know, when used unconsciously it's more likely to do the opposite. But placing alcohol in the same *category of intensity* as a drug like ayahuasca has helped me to reclassify what might be an appropriate use for it (more on this below)—adding a meaningful layer of caution any time my FOMA kicks in.

Speaking of spiritual healing, I also consulted Jessyka Winston, a Haitian Vodoun priestess and root worker who uses alcohol in her practice, regarding what she considers positive drinking. "Alcohol is part of my medicine, my spiritual practice, my magic, my culture and ancestral traditions, as well as part of my social life," she told me. "I associate with it as a spirit. If I overindulge, it will hand me my ass on a platter. If I respect this spirit and treat it with respect, it will elevate me. And as a spirit worker, I can tell you that applies to all spirits!" Jessyka grew up with alcoholism in her family and avoided booze socially until her early twenties. "The day I decided to drink," she told me, "I said to myself, 'I am in control.' I won't let fear have power over me where I avoid something like the plague because I believe I am inherently 'weak' to it."

Then there is the "prescription" that speaking coach Gail Larsen told me she once gave to a student who was finding it impossible to get past her nitpicking monkey mind and into the flow of delivering an authentic talk. "Drink two glasses of wine, get naked, and *then* plan your talk," Gail told her—the idea being that the booze (and the nudity) would shock her out of her head and tune her in to the language of her soul, the seat of our most effective storytelling.

Alcohol has been used medicinally throughout history—as an antiseptic and as an anesthetic in the early days of surgery, for example. In Chinese herbal medicine, it's sometimes used to invigorate *chi* and get the blood moving. "But you could go for a run and get the same effect, without the toxic rebound," says acupuncturist Sarah Emily Sajdak. She also explains that alcohol is generally used as a base to prepare an herbal mix, often by means of extraction, as opposed to taken neat.

These are just a few examples of the *reverence* it may be helpful to apply to your own view of booze. Personally, I now choose to think about alcohol as a Class A controlled substance, with the medicinal value of, say, morphine. Though the era of *actual* Prohibition (when the sale of alcohol was banned in the United States from 1920 through 1933) may have had mixed results, with an immediate drop in alcohol consumption followed by a booze-related crime wave and an increase in drinking among women and children (since all bets were off during that time as to what constituted "respectable" drinking), recategorizing alcohol for myself as a substance on a par with cocaine, LSD, or ayahuasca reflects for me how potent a drug it is.

Remember how during my last "reminder" I was blown away by how just *one sip* of beer could completely alter my mood? How it could paint my world in soft-focus shades of gold like pulling down the blinds on a raging hot day? And I used to down pints of the stuff! Learning to drink in the boozy landscape of the British magazine and nightlife scene, I quickly developed a very high tolerance. Not only has getting Sober Curious leveled that playing field, but having become more aware and appreciative of all my differing feeling states, I'm now hyperattuned to the impact of alcohol on my overall vibe—physical, mental, emotional, and spiritual.

Thanks to these developments, the kinds of questions I might ask myself if an opportunity for some seemingly positive drinking presents itself now are along the lines of:

How will this drink really *make me feel—now, in an hour, tomorrow morning?*

What am I drinking to not *feel?*

What is the purpose, the desired end result, of my drinking?

How many days will it take to recover the full extent of my faculties?

What else in my life may be affected as a result of this drink?

If you're used to cracking a bottle on autopilot, it can take years of practice to answer questions like these with integrity. Which, again, is where the gift of presence, as developed in your meditation practice, comes in. It's where you can begin to pay attention to the cues being communicated by your body, as you may have gotten used to doing on your yoga mat. Answering such questions truthfully (meaning in full integrity), and making relevant choices, also means getting really, really clear on what it means to be *you*—meaning what are *your* needs, *your* values, and *your* priorities—underneath all the social conditioning. And if you find yourself wrestling with questions like these for an undue amount of time (like, *obsessing* over them), then the answer to the question of whether or not to have that drink is very likely "not tonight."

I repeat. If you are going over and over and over the pros and cons of drinking at *X* family BBQ or *Y* birthday party, then the best advice I can give you is to stick with the abstinence until such "Sirens" have become background noise.

As you can imagine, approaching drinking this way *dramatically* decreases the situations in which it feels like an appropriate substance to partake of. I bet by now you're also sitting there wondering: Well what are these situations for *you*? And to be honest, I'm hesitant to share. I cannot state often enough that *your* Sober Curious path is yours and yours alone to walk. What I think of as "high-vibe" drinking may not even be in the same ballpark as what works (or doesn't) for you. As I've been writing this book, I've also, as noted, done a very convincing job of whittling down "as much as I like and as often as I want" to pretty much "nothing" and "none." But I do have one story that illustrates where I found myself a year or so ago.

Around this time, a Now Age podcaster named Noah Lampert had me on his show. He was super fascinated with the idea that getting Sober Curious could be part of a person's spiritual awakening, and he too wanted to know under what circumstances I used alcohol then. I told him how I'd recently had one and a half beers at a thexx concert. One beer because I was *curious* how it would make the music feel in my body, and an extra half because it felt *good*—but I also could tell that the pleasing effect would be reversed (meaning the numbing would set in) if I had any more.

Key to my choosing the beer on that occasion was the fact that the concert was happening in an open-air stadium encircled by trees, stars sprayed like glittering foam across the balmy night sky. "And there's something about the alchemy of loud music, nature, and a crowd of humans coming together to transcend, where alcohol triggers something *primal* in me," I told him. At which point he noted that I was essentially describing the kind of situation where my ancient pagan ancestors may also have used booze. Late summer, celebrating the harvest, overripe fruit beginning to ferment on the ground. Important to note: This was a couple of months *before* my Croatian New Moon rave experience, a Sober First during which I discovered that loud music, nature, and a crowd of humans coming together to transcend was enough to trigger my primal desire to dance like a banshee *sober*, too.

In fact, since then I've had similar experiences where even a few sips of a friend's beer at a concert or club has put a damper on my overall vibe. Made me kind of clumsy and my feet sort of heavy. Another clue as to what "as much as I like and as often as I want" looks like these days for me.

A Good Night Is Allowed to Look Different for Everyone

Another positive drinking experience might be staying Sober Curious while people all around you are getting trashed. The benefits of such an experience (that's an experience you can turn into an experiment) are threefold. Allow me to explain.

First, you get to witness up close and personal what the effects of drinking look like from the outside. IMPORTANT: This is not the same as propping up at the bar with your arms crossed judging these losers who used to be your friends for their idiotic drunken behavior. Nor is it the time to roll your eyes when somebody starts telling you the same story for the fourth time that night. Rather, contemplating the disconnect between how *elevated* being intoxicated can feel on the inside versus how . . . *basic* . . . it often appears in the real world has been eye-opening for me. It has helped confirm that my perception of myself as funnier, sharper, and sexier when drunk has more holes in it than a slice of Swiss cheese.

Not to mention that not judging the drunkenness of others can become a useful practice in and of itself—especially when applied to *you* and your own awkward/idiotic social behavior. For example, could choosing to see *yourself* through kinder, more loving, and less judgmental eyes have the same effect as three margaritas on the functioning of your right temporoparietal junction? Help you get more comfortable with just *letting it all hang out*, no numbing required?

Second, drinking *does* tend to loosen people up and make them laugh more—at least until the numbing and the repeating of the same fucking story four fucking times sets in. And being around people who are getting loose and laughing will likely mean *you* feel looser and find *yourself* laughing more, too, which is always a positive thing. Especially if it reminds you that you don't need the *substance* to loosen up and laugh—you just need to put yourself in a situation where getting loose and laughing is apparently the done thing.

Which is what happened on the Sober Curious bachelorette weekend I mentioned. At first, it was like my vibe synched with everybody else's buzz. The more Prosecco they popped, the more giggly and silly *I* got. When the cracks began to show and they went on to a club, well, that was when I took myself to bed. But the morning after, to my surprise, my "high" had lasted. I woke up feeling so much love for my friends and excited to hear what I'd missed. When they surfaced clutching sore heads, needing salty carbs, and, like, at least a decade of more sleep, we bonded again over what a great night we'd had. And to think—the friend whose party it was almost hadn't invited me, mainly because I don't drink. She had assumed that I wouldn't want to spend the weekend with a bunch of drunken idiots. The message? "A great night" can be a sober *or* a sozzled one.

Because, finally, being around other drinkers can be another reminder that, as I've said, *alcohol itself is not the bad guy.* It's the reasons we choose booze and the ways in which we use it that can become problematic. Also, being around other drinkers can remind you that you weren't making it up and that you're not crazy or deluded for thinking drinking *can* be fun—that it did, at some point, facilitate some joy and merriment. Let's not gaslight ourselves into denying point-blank the "good times"—even if, ultimately, they may have resulted in more bad times than they were worth.

Should you choose to go forth and mingle with the drinkers, remember that it's also not your place to dictate who's "got it under control" and who "needs" to get Sober Curious (or even Sober Sober). One person's toxic, addictive poison can be another person's ticket to party central. And that's okay. What counts is *your* integrity and your trust in *yourself* to make choices about booze (and everything else, for that matter) that align with *your* highest good.

With a couple caveats. If the people getting drunk in your vicinity are also becoming violent or abusive toward you (or anyone else for that matter), then remove yourself from the situation and feel free to report their behavior if appropriate. It's not your job to sit there and be a Sober Curious sponge to soak up any alcohol-related aggro. As for if you have a friend, a colleague, or a family member who is very obviously veering into risky drinking territory, know that your Sober Curiosity alone is modeling a positive example of nondrinking for them. Don't underestimate the impact of this. Eventually, they may follow in your footsteps. In more extreme cases, however, when a person's drinking is causing harm to themselves or others, see the list in the back of this book of organizations you can call for advice.

Which is also where the rubber meets the slippery, slippery road. Where we reach the extremely-tricky-to-navigate-while-wet intersection at which "positive drinking" ends and "problem drinking" begins. If you're still feeling in any way confused about the difference, then a positive drinking experience for you will always be the one where you don't.

What to Drink When You're Not Drinking

Whether you're out on an experimence with friends or are wondering about what to serve at a Sober Curious soiree of your own, these booze-free bevs are guaranteed to result in a positive drinking experience.

SODA AND LIME. Elevate the bog-standard by adding *lots and lots* of fresh lime, and maybe a couple of mint leaves and a pinch of salt. Invest in a SodaStream to make your own extra-fizzy soda water, and always serve it ice cold.

ALCOHOL-FREE BEER. Though controversial in Sober Sober circles (is the 0.5 percent alcohol content considered a relapse?), discovering the nonalcoholic version was a game changer for a former beer-drinker like me. Also packed with B vitamins, it's even used as an isotonic sports drink by members of the German Olympic ski team, who claim that it helps them train harder and recover faster. Europe is way ahead of the States when it comes to variety—look for imports of Clausthaler Dry Hopped, Erdinger Alkoholfrei, and BrewDog's Nanny State.

MOCKTAILS. You'd be surprised how excited bartenders get when faced with the challenge of mixing up some booze-free deliciousness. Tell them some of your favorite flavors and let them get creative. If you don't like sweet, an extra-spicy Virgin Mary delivers a grown-up kick.

KIN SOCIAL TONIC. Billed as "the conscious un-cocktail," this new alcohol-free party beverage is stacked with powerful adaptogenic herbs designed to bring about a subtle altered state. The desired effect is described as "focused relaxation" and generally kicks in after a single serving.

SEEDLIP. These alcohol-free spirits are made from various herbs and spices and come in three blends described as being "aromatic," "herbal," and "citrus." According to the company's website, its products are "based on the distilled nonalcoholic remedies from *The Art of Distillation* written in 1651."

TONIC + BITTERS. A more grown-up soda and lime that many a bartender has shared is also their booze-free go-to. Well, almost booze-free. The bitters contain alcohol, and so you should use only a drop or two.

CACAO. A true superfood, with more antioxidants than blueberries, plus tons of iron, calcium, and magnesium, raw cacao (derived from the same plant used to make chocolate) is said to boost cardiovascular health, thanks to high levels of flavonoids. The reason cacao powder is served in shots at Berlin raves, however, and taken in ceremonial healing circles, is for its heart-opening, mood-enhancing, and caffeine-like stimulating properties.

KAVA. The root of the kava plant makes a muscle-relaxing tea that tastes kind of like bitter mud (yum!) and is said to calm anxiety, stress, and restlessness and to be a sleep aid. Made strong enough, the tea also has a slight numbing effect on the mouth. It can be a good alternative to coffee for an after-dinner pick me up/chill me out!

WATERMELON JUICE. Literally just fresh watermelon blitzed in a high-speed blender with ice. Fucking delicious.

UNICORN SHOTS. I served these at a Club SÖDA NYC booze-free brunch, and you couldn't get people off the dance floor. Here's how you whip them up (makes eighteen shots):

- 4 cups almond milk
- 1 ripe banana
- 3 tablespoons raw cacao powder
- 2 tablespoons kava powder
- 2 tablespoons honey
- Handful of ice
- Rainbow cupcake sprinkles

Blitz all the ingredients except the sprinkles in a high-speed blender and divide among chilled shot glasses. Top with sprinkles and serve immediately.

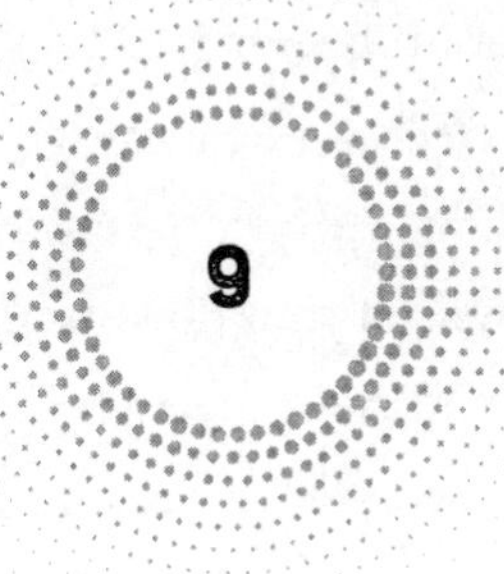

9

VISION FOR A HANGOVER-FREE SOCIETY

What would I do with my life if I were never hungover?

The majority of our time together thus far has been focused on unlearning the habit of drinking: getting to know the nature of the beast; considering the impact of getting Sober Curious on our relationships, our well-being, and our connection to our spirit; relearning how to feel our feelings; and discovering ways to get high on our own supply. Before I leave you, I want us to look up and out, to consider what might be the bigger picture when you commit to living Sober Curious.

One thing you already may have noticed is how many more hours there now appear to be in each day when you don't drink. How it feels like less of a *scramble* to meet deadlines and keep on top of your inbox. How the slivers of time outside work are positively yawning with potential, begging to be of service to a passion project or side-hustle. That whole weekends stretch out in front of you waiting to be filled with who-knows-what. If you're used to dedicating large chunks of your downtime to drinking, this sudden surplus of time and energy may even feel a little *intimidating*, could very well find you at something of a loss. What else even *is* there to do?

The things you find yourself doing instead of drinking may surprise you and may at first seem small and inconsequential. I'll always remember one Sober Curious friend telling me that it was only during sober stints that she could be "bothered" to put on body lotion after her morning shower. How this one small act of self-care became emblematic of the extra energy, commitment, and attention to detail that was seeping into other areas of her life—her work, her relationships, her well-being. The lotion was symbolic of the fact that more of her choices in general were coming from a place of self-love.

In my case, it was *needing* extra hours—not to mention energy, commitment, attention to detail, and self-love—to pour into my passion project, The Numinous, which kicked my own Sober Curiosity into high gear. For the first few years, I juggled freelance journalism with writing The Numinous, the former being how I paid my bills. There was no time *left* for being hungover. While I was trying to meet deadlines or stay committed to my passion project in the throes of a roiling hangover, regret for the productivity and enthusiasm I'd squandered on boozing weighed on me like a sodden wet blanket—a far cry from my experience of entering the workforce in my twenties, when hangovers were often kind of fun. Once I'd killed off the headache with a handful of max-strength painkillers, the blurry "who gives a fuck" mindset a hangover produced felt like an extended holiday from having to think too hard about life. Felt like something to bond over with my equally groggy desk buddies, swapping stories of the previous night's antics and still getting paid the same at the end of each month, regardless—no longer the case now that I was actively crafting an existence where the buck, literally, came to a screeching halt with me.

The medical term for "hangover" is "veisalgia"—coined in 2000 from the Norwegian word *kveis*, which means "uneasiness after debauchery," and the Greek word *algia*, or "pain," a combo I suspect you are familiar with. And why do hangovers get worse as we age? The simple answer is that the liver's capacity to process alcohol decreases. But personally, I believe our worsening hangovers are as much about *regret* as they are about biology. Since we've boozed long enough by now to know better, the "uneasiness" arises in the cognitive dissonance that comes from knowing that drinking is going to make us feel like shit, slack off, and potentially waste another day . . . and doing it anyway. This regret then mingles with the toxic chemical cocktail that's produced in our bodies when we drink.

For any seasoned drinker, how to cure a hangover is an age-old question—the only truly viable answer being, *Don't drink in the first place.* Kin Social Tonic cofounder Matt Cauble discovered this to be definitively the case when he set about trying to create a product that would allow people to drink but without the hellish aftereffects. Experimenting with various herbs, he discovered ways to combat the dehydration, acidity, and depletion of core nutrients that result from drinking. But the real issue, and one that cannot be reversed, is a compound called acetaldehyde, which is produced when the liver breaks down alcohol.

"It's the way our bodies process alcohol that does the most damage," says Cauble. It's also the closest medical science has gotten to explaining *why* we get hangovers (if the commonsense explanation that they are *a warning not to put this toxic poison into your body* isn't enough for you). According to Dr. Rachel Vreeman, an associate professor at the Indiana University School of Medicine, as quoted in an article for NBC News, "This toxin is probably the reason for a lot of the gross feelings that come with a hangover."

This discovery led Cauble to change tack. Since there can be no absolute cure for hangovers, what if he could create a drink that mimicked the desirable effects of alcohol—relaxation, optimism, and social ease, for example—but was toxin-free? After all, he notes, "many social rituals are based around some kind of ingestible, whether it's alcohol, tobacco, or tea." Meanwhile, with a background in the bio-hacking startup scene, and as is the case for many busy, single, city professionals, Cauble also noticed that in his circles "if you don't drink alcohol and go to bars, your options for socializing out of office hours can seem limited."

This is a sad state of affairs that any newly Sober Curious Netflix binger will no doubt be aware of. It can take time to rediscover the wholesome world of sober socializing, not to mention make new Sober Curious friends to enjoy it with. Ever the coffeeshop philosopher (literally, as we shall see), Cauble also sees this as a *potentially epic missed opportunity*. "When people are segregated out of social rituals (because they don't drink, for example)," Cauble explains, they're "missing out on so much creative potential. People don't act alone. It's when we come together that new ideas are alchemized, and change happens."

Not all drinking ends in bust-ups and blackouts, after all. For every night getting *wasted* (such an interesting use of terminology when discussing how to spend non-drunk and non-hungover hours!), I can also remember an impassioned debate fueled by Friday night camaraderie and firewater, the touch-paper of new ideas being lit in the intimate candlelight of cozy pub corners—even if there was usually very little in the way of follow-through. As for what would happen if more socializing happened sober? "With more options for people to commune without alcohol," Cauble says, "there will be more people contributing to the collective creative consciousness and furthering our understanding of each other and the world around us."

An insight that sparks more questions: *What am I here to contribute? What creative project is in me waiting to be birthed? How might my drinking or the dominant drinking culture be preventing me from pursuing this? What collective innovations are we sacrificing to our love affair with booze?*

To further illustrate his point, Cauble explains how the dawning of the European Enlightenment coincided with coffee from the Middle East infiltrating Europe and replacing alcohol as the dominant social stimulant. "Coffeehouses took over from bars as places people came together to discuss ideas about science and philosophy, and society took off," he says. Echoes of this change can be found in the rise of Silicon Valley—the drugs of choice among the tech nerds shaping our futures (for better or for worse) being cognitive function–enhancing "nootropics" and bio-hacking bulletproof coffee.

Beyond having more time and energy for our personal passions, the Big Question now becomes: *What would* we *do with* our *lives if* we *were never hungover?*

This question is all the more poignant when we take a look at the world around us—because, when it comes to politics, education, health care, the environment, economics, and social justice, never before has it felt more like the buck is coming to a screeching stop with *us*. As previously noted, faced with an uncertain future, we may have a collective sense that there is no more time to be *wasted*. That this is no time to be getting "out of it" and that we will all have to be very much *in it to win it*. Could one of the first brave steps be more and more of us choosing to live hangover-free?

More Hopeful, More Mentally and Emotionally Resilient

The "noble experiment" of Prohibition in the United States was undertaken as a response to the rising levels of alcoholism, domestic violence, and saloon-based political corruption of the time. But the resulting rise in violent gang and organized crime during Prohibition provides a neat illustration as to why criminalizing drug use, rather than treating the negative consequences of it (addiction, violence, mental and physical illness) as public health issues, doesn't work.

This is a subject for a whole other book (a good one being Johann Hari's 2015 *Chasing the Scream: The First and Last Days of the War on Drugs*, if you're interested). The reason I mention it here, however, is that my *vision for a hangover-free society* does not look like a society in which it is illegal to consume alcohol. The whole point of getting Sober Curious is to come to your own conclusions about the true effects of booze and to then apply this heightened awareness to making conscious and informed choices about the role it plays in your life.

I believe Prohibition didn't work because making a drug illegal doesn't do anything to address or rectify the underlying reasons it's being used in the first place. As I've discussed, I hope that getting in touch with *why* you drink will be an integral part of your own Sober Curiosity (with professional help if needed)—whether you identify as being kind-of-just-a-little-bit-addicted or are an occasional drinker concerned with the wider impact of our collective numbing with booze.

In the case of the former, this may mean investigating what you are trying to escape through your drinking—the root of your discomfort with simply being you and the painful emotions anchoring you to past traumas or anxiously constructing possible future scenarios. In *The Life-Changing Magic of Tidying Up*, Marie Kondo states, "When we delve into the reasons for why we can't let something go, there are only two: an attachment to the past or a fear of the future." And the same applies to alcohol.

As you choose to remove the booze and live in the here and now, perhaps you will seek some sort of therapy or other healing modality to help you sift through the emotional backlog. Maybe you'll instigate a few difficult conversations (real or imaginary; I find writing letters and emails I will never send to be very therapeutic) with the perceived "perpetrators" of your existential pain—the aim of this being to make peace with your past, thus freeing you to imagine a future based on the hopes and desires of the person you are today, versus whichever version of you keeps being triggered.

In *The Biology of Desire*, Marc Lewis claims that "this process of reflection and perspective taking was precisely what helped [two of the addicts profiled] to overcome the now appeal of drugs"—"now appeal" being what Lewis defines as "the tendency for humans . . . to value immediate rewards over long-term benefits" and a key component in developing addictive behaviors. He goes on to explain how "what kept [alcoholic] Johnny sober was the insight he gained about who he was, who he'd been as a child, and who he could still hope to be without the balm of alcohol." *Insight that would shape a possible future it was worth both waiting and working for.*

"Worth waiting and working for" because all the above takes time, effort, and a willingness to confront demons that may at first be cloaked in shadow and will require the support of family, friends, and perhaps a larger community of similarly minded seekers. In certain cases, it may also mean a financial investment in some kind of therapy. But I believe that a mass healing of our inherited emotional pain, individually and as a collective, could have a ripple effect of epic proportions. And considering that alcohol is the most commonly used substance for "escaping" our problems, I also believe that this healing will be possible only when we're ready to rip off the boozy Band-Aid and tend to the wounds underneath it.

Why a ripple effect? As I've said before, the common "side effects" of quitting drinking and then tackling head-on whatever personal "issues" tend to arise may include more energy; more confidence; a more optimistic outlook; trust in yourself and your decision-making process; better relationships; less anger, boredom, and frustration; and *a desire to contribute something useful and to help others.*

It kind of goes without saying (but I'll say it anyway) that a hangover-free society would be a physically healthier society, what with alcohol being the third leading cause of preventable deaths in the United States. There would be less heart disease, less cancer, less obesity, less Alzheimer's. Less clogging-up and financial taxing of emergency rooms with Friday night "disco damage." We would live longer, too, the latest research from The University of Cambridge showing that even one glass of wine per day can shorten a person's life by anything from six months to five years.

But what would be the point of a long and healthy existence if it were a miserable and self-absorbed one? To me, *a more mentally and emotionally resilient society* is by far the most exciting and potentially important consequence of a mass Sober Curious awakening—more of us existing at *a higher vibrational state,* and this informing every thought we think, word we speak, and action that we take. Informing every choice that we make—from the food we eat and the work we do, to the creative projects we pursue, and the causes to which we might contribute our surplus time, money, energy, and other resources.

All of which begins with "cleaning up your side of the street"—a phrase that's used in recovery circles and yoga communities alike to mean rolling up your sleeves and getting your hands dirty fixing your own shit, versus dumping all your crap in your neighbor's backyard and expecting somebody else to clear it up. The problems (large or small) that we're often drinking to distance ourselves from? The temptation sometimes can be to fix blame and make them all external to us, a result of something "they" said or did. Not to take away from the wrongs we may have been subjected to, or to make any traumatic experiences *our fault*, but it's only when we take responsibility for how we respond that we truly become free to move forward with dignity.

It's Not Them, It's You

"Other people seem less annoying." I think I laughed out loud when I read this in the list of benefits of a decent night's sleep from the US Department of Health and Human Services. But I've found it to be 100 percent true, as well as a surprise plus that can be applied to getting Sober Curious in general.

Why? Because having more patience and feeling less frustrated with life and the other people in it is a direct result of taking responsibility for your own experience. Beginning to trust that whatever shitty circumstances you may find yourself faced with, how you choose to respond—including how much time and energy you spend worrying or feeling pissed-off about things—is always in your power. When I found myself disillusioned with the magazine career I'd worked so hard for, my first instinct was to blame the shallow, competitive nature of the industry. But getting Sober Curious helped me see that I'd been drawn to the glamorous world of fashion and celebrity as a result of my own lack of self-worth. If I wanted to create a life that was more meaningful to me, I would first have to address the root of these issues.

As for how this relates to drinking? It comes back to the role of our emotions—how we *feel* about something and how this informs our first responses to external events, letting our minds know which thoughts to generate, and thus informing our bodies about the right next actions to take. When these three steps align—how we feel, how we think, and how we act—we're able to live with integrity, meaning living in alignment with our personal values and needs. This then leads to cognitive resonance and a feeling of inner peace.

When we choose to manipulate the majority of our feeling states—from joy and relaxation to disappointment and outrage—with booze, they become disconnected from whatever gave rise to them in the first place. We use a glass of wine to paper over the cracks of a bitchy, gossipy friendship here, a beer with the boys to allay feelings of inadequacy or a fear of emotional intimacy there. This inebriating buffer on our feelings separates us from our power either to pursue more of what feels genuinely joyful or to use our naturally occurring indignation to confront issues head-on with integrity and awareness.

Which, in the case of those annoying other people (or situations), doesn't take away from what may have been said or done, but it does allow for a reconciliation of sorts, even if it's internally, before the situation gets blown out of proportion—before the disconnect between how you feel and how you're acting begins to eat at you day in and day out, weighing heavier and heavier on your spirit, making it harder to muster the energy to take empowered action in *any* area of your life. Like in that 1975 kids' storybook *There's No Such Thing as a Dragon*—where a boy wakes up one day to find a baby dragon in his room. Ridiculous, his mom insists. After all, "There's no such thing as a dragon!" But the longer everyone refuses to acknowledge the dragon, the bigger it grows—until it's filling every room of the house.

As you survey the landscape of your newly Sober Curious life, you may find some outsized dragons of your own lurking in the undergrowth. Perhaps some have even taken up residence front and center in your life. Considering how *hungry* dragons get, the vast reserves of energy it takes to sustain just *one* of these demanding beasts, it's no wonder you feel so *tired* all the time. And yes, now is the time to look them in the eye and show them the door. A process that may, initially, seem as inviting as a bungee jump over a shark tank—but one that ultimately is an essential part of creating a long and healthy life of sustainable ease and peace of mind, an existence infused with unexpected bursts of joy.

In *Daring Greatly: How the Courage to Be Vulnerable Transforms the Way We Live, Love, Parent, and Lead* (2012), Brené Brown frames drinking as a way of letting ourselves off the hook from *daring* to be truly *great*. Any time we drink, she says, we're numbing whatever emotion is arising to inspire our next right, if vulnerable, action. That "edge" we're so keen to rub smooth with a "relaxing" glass of Pinot? Precisely the place where we have the opportunity to lean in to the full force of our feelings, however unwelcome or uncomfortable, and use them to propel ourselves into dazzling, unchartered territory. The place where endless, if at first daunting, possibilities await.

The other reason people seem less annoying to us when we get Sober Curious? Because we become more discerning about who we spend our time with. Now, with all these newfound possibilities to pursue, comes an appreciation for the preciousness of each and every hour of each and every day—a realization that there's no time to waste on whiners (accompanied, I hope, by an influx of high-vibe individuals who accept you fully as you are and reflect what's truly important to you).

Which will also free up yet more energy that can be funneled into the *desire to contribute something useful and to help others* that I have mentioned. It's one thing to keep our own side of the street looking nice and neat, but how about the wider world, where oppression, inequality, and suffering are a seemingly insurmountable daily reality for far too many people? As we tend to the by-no-means-insignificant task of taming our dragons or showing them the door, it's highly likely that we will also find ourselves asking about our role in the epic evolutionary shift we now face. In which case, please join me for a moment down here in the streets of New York City.

A Sober Curious Society That's Better for Everybody

In the Bronx, specifically—a predominantly black and brown borough of New York that happens to be the most racially diverse area of the United States and where the average household income is 34 percent lower than the national average. It's also the birthplace and stomping ground of Hawk Newsome, president of Black Lives Matter of Greater New York.

I first met Newsome when he came to speak at a Club SÖDA NYC event. Having quit drinking two years previously, he was taking part in a panel on the role of Sober Curiosity in building conscious communities, since living hangover-free had had a hugely positive impact on him and his activism—not least because he became so much more productive. "Waking up hungover," he told me, "it was all I could do to get up and put my kid on the bus to school. I'd spend most of my day just killing time—they don't call it getting 'wasted' for nothing." We met again in downtown Manhattan, as he was scheduled for an interview with *Newsweek* at the magazine's nearby office. He went on to tell me about the rest of his day.

"When I got up I played with my daughter for a while, then did some yoga and a few pushups. I went to get a letter notarized. Sat down for a meeting with Dr. Mark Hyman, a superstar in functional medicine. I made an appointment with the sheriff's department. Then I took a phone call with a victim who needed my assistance. She was shot five times by her partner. Just think about that dynamic . . ." he trailed off. "I was an abuser, and now I'm fighting to get justice for a woman who was shot by her former partner . . . thank God I'm able to help, right?"

Newsome is outspoken about the fact that he also got aggressive when he drank, mainly with the people he was closest to. "Every time I'd get into a fight, it was alcohol related. I would go around ranting—berating and belittling people. Ultimately, I wasn't happy with myself and I thought I could be doing more [as an activist and leader], but I would just drink to mask it." It's not something he's proud of, and he shares this part of his story as another example of the positive impact that sobriety has had on his life. And could have on so many lives.

Especially since he also points out: "I'm passive in a lot of ways, and people take advantage of that. I didn't love myself enough to tell you exactly how I felt in the moment. The things that I let slide would resurface when I was drunk." Sound familiar?

It could be said that many of us have been engaged in a largely passive acceptance of things that are *not okay*. Political corruption. Environmental destruction. A broken health-care system. Could it be our subconscious anger and shame about the things *we* let slide, as a society, is part of what we're masking when we drink? That the violence and pain of our hangovers are a result of internalizing this anger and turning it on ourselves?

Here is where I must acknowledge that it's fine for me, a financially stable, educated white woman, to tell you that, without question, things will improve *dramatically* when you get Sober Curious; that the possibilities for what to do with all those extra non-hungover hours are *endless*. This may seem far-fetched, if not downright ignorant, if you're living on minimum wage, and have inherited the legacy of generations of systemic racial oppression, or both. Or have grown up in a community that is routinely terrorized by authorities and/or violent crime; have lived the kind of life that a person like Hawk Newsome has dedicated his non-hungover days to lobbying for and protecting.

Given the "now appeal" and comforting embrace of liquor, it's not at all surprising that risky drinking is rising fastest in the most marginalized communities.

Which is to say nothing of the way that, as Aaron Rose puts it, "institutionalized racism [for example] has also benefited from the ways in which alcohol and drugs numb us, dim our memories, and reduce our capacity for connection and critical thinking." He claims that "by choosing sobriety, we reclaim our full awareness [of our own roles in perhaps perpetuating these systems of oppression] and our humanity."

Newsome believes that the positive possibilities for *everybody* increase exponentially when booze is removed from the equation. On a very basic level, he says, "discovering you can still enjoy your life without alcohol is proof that you are not a victim and that nothing else is in control of you"—meaning living, joyous proof that we all have the power to generate our own happiness from within, regardless of external pressures. To which I will add, as long as we also have meaningful human connections in our lives and the sense of self-worth these generate. This is another reason to thank Goddess for the free community support of an organization like AA—as well as *the influx of high-vibe individuals who accept you fully as you are and reflect what's truly important to you* that is often the result of a person's sustained Sober Curiosity.

Most important, however, Newsome's path has shown him that not numbing out with alcohol is "how we learn to address our problems instead of trying to run from them." This is a super-empowering message wherever you land on the spectrum of privilege, especially considering the point of global crisis where we now find ourselves. When it comes to addressing the most complex of our collective issues, what would it mean to take *a more sober view* of the path ahead?

Radical Dharma: Talking Race, Love, and Liberation (2016), by African American Buddhist teachers Rev. angel Kyodo williams and Lama Rod Owens, as well as academic Jasmine Syedullah, sets out a manifesto for transforming the chains of the us/them, master/slave mentality that they believe are at the root of *all* inequality in the world—racial, social, gendered, and economic. Inequalities that imprison us within a system built to dehumanize us all. When it comes to dismantling these forces of oppression, Rev. angel writes in the book's introduction, "we have to be the transformation . . . The only way we can do that is to observe the construct that we're in [in this case "white supremacist capitalist patriarchy," as coined by author and social activist bell hooks] instead of trying to tinker with it right away with the same blind spots that we came to the problem with." And the only way to do *that* is with absolute, undiluted clarity and by bringing to it, fearlessly, the full presence of our awareness.

I'm highlighting institutionalized racism here because if you get deep into the weeds of sexism, homophobia, economic inequalities, global political warmongering, the refugee crisis, environmental issues (name ANY of our Now Age woes), you'll find that the roots are firmly entwined with those of our white supremacist economic system that only benefits from white people's unwillingness to examine our (often unconscious) roles in perpetuating it.

Confronting my part in this is as dark a piece of shadow work as I have been asked to face as a result of my Sober Curiosity, and unraveling its implications, for myself and as a society, "is a process a lifetime in the making," Aaron Rose says, "requiring layer after layer of inquiry and choice after choice to reprogram our minds to see the world in a new way." This is an endeavor that we must be fearless in undertaking, he continues, because "individual and collective trauma, compounded over generations, can make it very painful to sit with the full reality of how we and others feel." In this instance, the pain and suffering of the oppressed (people of color), and the fear, guilt, and shame of my whiteness alone positioning me in the role of "oppressor." However, he adds, "I see an awakening to Sober Curiosity as fundamentally a return to greater presence with ourselves and others. The initial reckoning can be extremely hard, but once we face what we've been ignoring, true peace and love are right around the corner."

Let's pause with this for a moment. Take a breath.

Part of me wants to apologize for getting "heavy" with you in this chapter and is super uncomfortable about stirring this stinking pot in a book that was supposed to promise "joy." But this is also a book about *spirit* and about *healing*, and as Rev. angel says in *Radical Dharma*, "If you're going to any place of spiritual enrichment in which you are not meaningfully experiencing discomfort, not all the time, but meaningfully uncomfortable frequently, you are not doing your work, and you are not walking the path of liberation." And if drinking is about "escape," then getting Sober Curious means I for one am ready to graduate to seeking ultimate liberation.

Regaining (Collective) Consciousness One Sober First at a Time

> *"In these days . . . there is no greater need for sober-thinking, healthy debate, creative dissent and enlightened discussion."*
>
> —MARTIN LUTHER KING JR.

Whatever may be stirring in you as you contemplate living hangover-free, one thing is certain: When it comes to transforming the world we live in, at both the personal and collective levels, we need to be fully conscious to do it. To have *all of our wits about us.* I'm not suggesting that more of us quitting drinking is *the* answer to all the world's problems, but how do we raise our level of consciousness, or awareness? The better to fully understand where we are at? Step one may as well be to remove a known depressant that is routinely pulling our energetic vibrations down to a lower state. That is preventing us from experiencing high-vibe states of acceptance, reason, and willingness, for example, by keeping us down in the trenches of anger, fear, and apathy. After all, as Matt Cauble reminded me: "When you drink, you become more hopeful and optimistic because your problems appear to go away. But this optimism is literally the result of *ignorance.* To create positive change, we need to be hopeful *and* fully conscious."

I found a YouTube clip where Eckhart Tolle, author of the self-help classic *The Power of Now,* speaks specifically to the detrimental role of alcohol in cultivating presence—the only place from which we can take empowered, fully-aligned action. In it he claims that: "[Consuming alcohol] can represent a significant setback in the arising of consciousness. If awareness, presence, is arriving into you . . . it's like you're going up the escalator, and drink will knock you right back down again. The moment you have the first drink, awareness diminishes. You have the second drink, it diminishes even more. With the third drink it's gone completely."

This metaphor helps explain why a state of heightened awareness is the inevitable result of sustained periods of abstinence—during which you keep on going "up the escalator." A level of awareness that may be downright uncomfortable at times, but which can eventually be applied to any and all setbacks we find ourselves faced with—from rewiring a tired old drinking habit to ending a dysfunctional relationship or work situation. Perhaps, even, to playing an active part in radically transforming a broken political and economic system that requires the oppression and exploitation of some people, and our planet, in order for a privileged few to profit.

This is another window into the *really* Big Picture ramifications for a hangover-free society, along with the painful truths it becomes harder to ignore when we're choosing to meet each and every day with utter clarity and from a dizzying vantage point that's as invigorating as it can be vertigo-inducing. A perspective that may begin to raise more Big Questions about your role in the future thriving of our human family.

> ***In what ways has my numbing made me complicit in propagating forces of oppression? What responsibilities does being human come with? Where have I been giving away my power? How can I shift my perspective so I am able to see past my own blind spots to see the whole picture?***

I'm not suggesting that getting Sober Curious will make you an activist overnight or an expert practitioner of the kind of "radical dharma" explored in the book of the same name. But it is possible to switch from complicity and apathy to engagement and accountability in a heartbeat—the time it takes to release ignorance and choose *radical consciousness.*

And then there's that *ripple effect*—the positive impact you begin to have on others and the world around you when you begin to focus less on the petty annoyances of life (office gossip, social media trolls, *hangovers*) and more on what you may have let slide as a result of leaking time and energy on those things—thus making space for more of what brings you joy. More and more, research in physics—quantum mechanics, string theory, entanglement theory—is showing how everything in the universe is interrelated. As Sah D'Simone says, "On some level, we're all completely interconnected, and so what we do and don't do matters. If our thoughts and actions are coming from a happy, healthy, calm, and productive place, then this is having an impact on every other being we come into contact with."

Hawk Newsome points out that "when you're so focused on numbing your pain or escaping the situation that you're in, it's impossible to focus on helping other people." Booze narrows our focus even further. Given its physical, emotional, and spiritually toxic side-effects, and how it alleviates our pain by suppressing the very emotions that are trying to guide us to higher ground, it keeps us locked in a cycle of suffering. Alcohol exacerbates existing problems and creates more, in the form of hangovers, cancer-causing toxic reactions, and morning-after regret. All this makes us more self-absorbed, more overwhelmed by the enormity of the problems we face, and even less likely to believe in our ability to contribute anything of value.

Perhaps your contribution will simply be spending more time with and having more energy for family and friends in need. Maybe you'll have a more inspiring social media feed. Maybe you'll cultivate a less reactive mindset and find more room in your life for creativity. And, just perhaps, when enough of us begin to engage in these small shifts in consciousness and behavior, they will begin to have an impact on the Big Picture, too.

For in our everyday lives, this is still a story of us down here on the ground, trying to do our best with what we've been given, choice by choice, and step by step. This might include a shitty role in the chorus line that we got assigned at birth, or a boozestory spiked with pain that we inherited from our parents, our peers, our society. But when it comes to rewriting the script, you should never underestimate the impact of simply choosing to live hangover-free. With an open mind and a willingness to go against the grain, I believe you will soon come to agree with Holly Whitaker when she says that "removing alcohol from your life is not about asceticism or deprivation. It is removing a shackle most of us don't really see so that we can live into our full potential."

For many of us, getting Sober Curious begins with a simple question: *Would my life be better without alcohol?* To discover the answer for yourself, all that remains is to put the cork back in the bottle, open your eyes, and see.

AN ALTERNATIVE 12 STEPS FOR LIVING SOBER CURIOUS

Getting Sober Curious is not just for "dry January." This is a lifestyle—one that will transform every area of your existence. Some of these shifts will feel like winning the lottery. Some will make you question the nature of being human. Taking your Sober Curiosity seriously means showing up as willingly for the challenges as you do for the surprisingly joyful parts, though. And one thing is certain: Life is about to get *way* more interesting.

To keep you on track as you take what you've learned here out into the big, wide, often fairly wasted world, here are my "Alternative 12 Steps" for living Sober Curious.

1. Choose abstinence.

It's never too early to choose not to use booze. You don't have to hit rock-bottom. You don't have to get sick. Remember that on the other side of this life-enhancing choice is another day hangover free, another opportunity to grow your innate confidence and to feel more and more like you.

2. Just say no to moderation.

Moderate or "mindful" drinking is still drinking, with all the same side effects that led you to get Sober Curious in the first place. Attempts to moderate will always lead you back to square one. Remember how hard it is to change a habit. The only way to get unaddicted to alcohol is to *stop drinking alcohol.*

3. Know your triggers.

Pay attention to when FOMA kicks in and question all the reasons why. Befriend the part of yourself that craves a drink in certain situations. Pay attention to how quickly the desire begins to disperse when you question it. In moments of vivid euphoric recall, play the movie forward, all the way to the bitter, slurring, end.

4. Embrace Sober Firsts.

See every Sober First as an opportunity to remind yourself how happy, confident, and relaxed you are without alcohol. Even the ones that are about as fun as sticking hot needles in your eyes—since these will shine a light on who your friends are *not*, and the situations where you *do* feel happy, confident, and relaxed. You will NEVER regret not drinking the morning after.

5. Don't make booze the bad guy.

Since the only way booze can do you harm is if you drink it, there's no reason to move through life trying to avoid alcohol like the plague. It's the reasons why and the ways in which we use booze that can lock us into a bum deal with the devil. Commit to uncovering your "whys." Trust yourself not to imbibe.

6. Don't judge others or preach.

Your Sober Curious journey is yours and yours alone. It is not your job to be the judge of anybody else's drinking habits. Just do you. Focus on living your own magnificent Sober Curious life and being a positive role model with the choices that you make. "Act natural" about not drinking, and chances are you'll naturally attract others to this path.

7. Be grateful for the reminders.

Whether it's a carefully orchestrated experiment or an unexpected slip, every time you drink can be a reminder of why you don't. No matter how much fun you have, this will be followed by some un-fun. It's inevitable. As the reminders come fewer and farther between, the "fun" part will feel less and less like it's worth what's waiting on the reverse.

8. Be honest.

With yourself, with others, and in everything that you do. About drinking. About how you're feeling. About what makes you happy, mad, sad, and everything in between. Speak up when you see an injustice. And when you experience an unexpected moment of joy! Align your thoughts, words, and actions. Living with honesty and integrity will lead to cognitive resonance.

9. Find your Sober Curious crew.

Your crew are the people who love, accept, and embrace you as your sober self. They don't have to be Sober Curious too, but they will respect your choice. They will not ask endless questions about why you're doing this (unless you want to talk about it). They will not make it all about them. And they will be your sober buddies on Sober Firsts that feel too daunting to do solo.

10. Feel your feelings.

But don't be overwhelmed by them. Your feelings are not YOU; they are messengers from your soul about the next right actions to take. Throughout the day, repeat to yourself, "I notice that I'm feeling *X*." Observe what *X* is a response to. Resist the desire to react, to suppress the feeling, or to run from it. If in doubt, find someone to help you talk it out.

11. Replace spirits with spirit.

Find a meditation practice that works for you. This is how you get spirit on speed dial, the connection you were possibly seeking in spirits. Try a yoga class. Find a way to express your creativity. Spend time in nature. Lose yourself in music. Dance with abandon (alone is awesome). These are all ways to bring spirit into your life that will fill your cup until it is overflowing.

12. Remember the Big Picture.

At the beginning or end of each day, express gratitude for the fact that you get to be alive and for the things that make life worth living. Notice how your world is getting bigger and bigger. How you are spending less time focused on your problems and have more energy to channel into your unique contribution. See your life as an act of service. Be brave. Believe that you can make a difference.

ACKNOWLEDGMENTS

First, I acknowledge all the individuals courageously walking the Sober Curious path, for questioning the status quo and for boldly daring to come to your own conclusions about booze. I also salute those working their recovery from any and all addictions, who, in doing so, are helping create a happier, healthier society for all.

I would not have been able to write this book (meaning, would not even have considered the possibility of *me* writing a book like *this*) without the love and support of: Mark Herman; Rana Reeves; my original SÖDAs (Kate, Kerrilyn, Tatum, Kirsty, Karin, and Vix); Jenna and her hens; my Club SÖDA NYC cofounder, Biet Simkin; and all our SÖDA guest speakers and the venues that have hosted us.

For the wisdom, insights, and time you contributed to this book, I extend my heartfelt gratitude to: Aaron Rose, Alexandra Roxo, Hawk Newsome, Holly Whitaker, Jen Batchelor, Latham Thomas, Jessyka Winston, Marc Lewis, Matt Cauble, Sah D'Simone, Sarah Emily Sajdak, Shaman Durek, Shona Vertue, Stephanie Snyder, and Tommy Rosen.

And for the ongoing inspiration, thank you: Fat Tony, Gabby Bernstein, Glennon Doyle Melton, Josh Korda, Laura Willoughby, Luke Simon, Mia Mancuso, Russell Brand, Sapphia

Haddouche, Tawny Lara, and Diego Perez. Thanks to you, sober is definitely cooler!

On the publishing side, Coleen O'Shea, your enthusiasm and guidance has been a godsend. Onward! Team HarperCollins, thank you for giving me another go-around. And Libby Edelson, your sensitive, no-nonsense, and whip-smart edits have, again, been the making of this manuscript.

I also acknowledge my parents, Nancy and Paul, for teaching me to know my own mind, to trust, and speak, my own truths, and for always allowing me to make my own choices about my own life. This book is a result of the values you instilled in me.

And finally, Simon Warrington, thank you for hearing me out on this one. I love you.

RESOURCES

Alcoholics Anonymous offers free 12-step recovery support groups globally: https://www.aa.org/.

Club SÖDA NYC is a Sober Curious event series from Ruby Warrington and Biet Simkin: https://www.clubsoda.nyc/.

Club Soda UK offers sober resources, community, and meetups in the United Kingdom: https://joinclubsoda.co.uk/.

Dharma Punx NYC offers Buddhist tools for overcoming addiction in weekly dharma talks that can be accessed online: https://www.dharmapunxnyc.com/.

Federally funded health centers often offer mental health services. Find one near you at https://findahealthcenter.hrsa.gov/.

No Beers? Who Cares! is a sober membership program based out of New Zealand: https://nobeerswhocares.com/.

Open Path Collective is a network of registered therapists offering sessions priced $30 to $50: https://openpathcollective.org/.

Recovery 2.0 offers tools for healing and connection to help people thrive in recovery: https://recovery2point0.com/.

She Recovers is a women-only community for women recovering from all kinds of addictions: https://sherecovers.co/.

Smart Recovery offers self-help tools for overcoming addiction and hosts meetings in the US, Canada, Australia, Denmark, Ireland, and the UK: https://www.smartrecovery.org/.

Talkspace offers daily online and text therapy with licensed therapists from $49 per week: https://www.talkspace.com.

Tempest is a sobriety school and group sober coaching program from Holly Whitaker: https://www.jointempest.com/.

NOTES

INTRODUCTION

p. 9: *the latest figures suggest that up to* one in eight *Americans is dependent on alcohol:* Marc A. Schuckit, "Remarkable Increases in Alcohol Use Disorders," *JAMA Psychiatry* 74, no. 9 (2017): 869–70, https://jamanetwork.com/journals/jamapsychiatry/article-abstract/2647075.

CHAPTER 1

p. 18: *"preoccupation with substance use":* "Public Policy Statement: Definition of Addiction," American Society of Addiction Medicine, April 12, 2011, https://www.asam.org/resources/definition-of-addiction.

p. 21: *Drinks company Pernod Ricard alone invested $421 million in advertising in the United States in 2017:* "Pernod Ricard Company's Advertising Spending in the United States from 2013 to 2017," Statista, https://www.statista.com/statistics/463837/pernod-ricard-ad-spend-usa/, accessed August 14, 2018.

p. 21: *according to a 2007 report published in the medical journal* The Lancet: Mike Nudelman and Eris Brodwin, "Alcohol Is One of the Five Most Addictive Substances on the Planet," *Independent*, October 12, 2017, https://www.independent.co.uk/life-style/most-addictive-substances-alcohol-nicotine-cocaine-barbiturates-the-lancet-a7996616.html.

p. 23: *only hospitality and construction work see higher rates of substance use disorders*: Donna M. Bush and Rachel N. Lipari, "Substance Use and Substance Use Disorder by Industry," *CBHSQ Report*, April 16, 2015, https://www.samhsa.gov/data/sites/default/files/report_1959/ShortReport-1959.pdf.

p. 34: *between 40 and 60 percent of people in abstinence-based recovery programs relapse:* Ruben Castaneda, "Why Do Alcoholics and Addicts Relapse So Often?," *U.S. News & World Report*, April 24, 2017, https://health.usnews.com/wellness/articles/2017-04-24/why-do-alcoholics-and-addicts-relapse-so-often.

CHAPTER 2

p. 44: *it's when alcohol becomes a friend, the thing that we turn to to relieve our stress, to numb our pain:* Deidra Roach quoted in *Risky Drinking*, 2016, https://www.hbo.com/documentaries/risky-drinking.

CHAPTER 3

p. 70: *drinking is the leading predictor of a person developing sexual dysfunction:* Bijil Simon Arackal and Vivek Benegal, "Prevalence of Sexual Dysfunction in Male Subjects with Alcohol Dependence," *Indian Journal of Psychiatry*, 49, no. 2 (2007): 109–12, https://www.ncbi.nlm.nih.gov/pmc/articles/PMC2917074/.

p. 70: *women who binge drink (five drinks or more at one time) are five times more likely to contract gonorrhea: Alcoholism: Clinical & Experimental Research*, "Women Who Binge Drink at Greater Risk of Unsafe Sex and Sexually Transmitted Disease," ScienceDaily, September 15, 2008, https://www.sciencedaily.com/releases/2008/09/080904215613.htm.

p. 74: *89 percent of "incapacitated" sexual assault victims said they had been drinking alcohol or were drunk before they were assaulted:* Christopher P. Krebs, Christine H. Lindquist, Tara D. Warner, Bonnie S. Fisher, and Sandra L. Martin, "The Campus Sexual Assault (CSA) Study: Final Report," October 2007, https://www.ncjrs.gov/pdffiles1/nij/grants/221153.pdf; Jessie Baimert, "Kasich on Sexual Assault: Don't Go to Parties with Alcohol," *Enquirer*, April 15, 2016, Cincinnati.com, https://www.cincinnati.com/story/news/politics/blogs/2016/04/15/kasich-sexual-assault-dont-go-parties-alcohol/83084604/.

p. 74: *"mouthy, up for a laugh, took her clothes off and could out-do any male companion in the drinking stakes":* Louise Donovan, "The Rise and Fall of the Ladette," Vice, International Women's Day, March 8, 2017, https://www.vice.com/en_uk/article/ypkp9m/the-rise-of-fall-of-the-ladette.

CHAPTER 5

p. 104: *the number of teetotal sixteen- to twenty-four-year-olds increased by over 40 percent between 2005 and 2013:* Oscar Quine, "Generation Abstemious: More and More Young People Are Shunning Alcohol," *Independent*, January 15, 2016, https://www.independent.co.uk/life-style/food-and-drink/features/generation-abstemious-more-and-more-young-people-are-shunning-alcohol-a6811186.html.

p. 104: *the British National Health Service is considering the use of "drunk tanks":* "NHS to Consider Routine Use of 'Drunk Tanks' to Ease Pressure on A&Es," *NHS England*, December 29, 2017, https://www.england.nhs.uk/2017/12/drunk-tanks/.

p. 105: *alcohol use is on the rise, up 11 percent overall between 2002 and 2013:* Bridget F. Grant, S. Patricia Cou, and Tulshi D. Saha, "Prevalence of 12-Month Alcohol Use, High-Risk Drinking, and *DSM-IV* Alcohol Use Disorder in the United States, 2001–2002 to 2012–2013," *JAMA Psychiatry* 74, no. 9 (2017): 911–23, https://jamanetwork.com/journals/jamapsychiatry/fullarticle/2647079.

p. 121: *According to the US Department of Health and Human Services, routine deep sleep leads to:* "Get Enough Sleep. The Basics: Health Benefits," U.S. Department of Health and Human Services, healthfinder.gov, https://healthfinder.gov/HealthTopics/Category/everyday-healthy-living/mental-health-and-relationship/get-enough-sleep#the-basics_2, last updated August 14, 2018.

CHAPTER 6

p. 133: *77 percent of people interviewed living in NYC said they are "stressed":* James Manning, "The Time Out City Life Index 2018," *Time Out*, June 13, 2018, https://www.timeout.com/things-to-do/city-life-index.

p. 136: *helps foster mental and emotional resilience, improve self-esteem, and boost empathy:* Amy Morin, "7 Scientifically Proven Benefits of Gratitude That Will Motivate You to Give Thanks Year-Round," *Forbes*, November 23, 2014, https://www.forbes.com/sites/amymorin/2014/11/23/7-scientifically-proven-benefits-of-gratitude-that-will-motivate-you-to-give-thanks-year-round/#ac5946c183c0.

CHAPTER 7

p. 152: *gave feeling states a numeric value in what he called his Map of Consciousness:* David Hawkins, "A Clear Map to Your Spiritual Enlightenment," Heal Your Life, July 14, 2015, https://www.healyourlife.com/a-clear-map-to-your-spiritual-enlightenment; "Dr. Hawkins," Veritas Publishing, https://veritaspub.com/dr-hawkins/, accessed August 14, 2018.

p. 168: *that research has shown helps people stay sober longer:* Jill Suttie, "Can Helping Others Keep You Sober?," *Greater Good Magazine*, April 14, 2016, https://greatergood.berkeley.edu/article/item/can_helping_others_keep_you_sober.

CHAPTER 8

p. 179: *(according to the National Institute on Alcohol Abuse and Alcoholism) 90 percent of recovering alcoholics "relapse":* "Alcohol Rehab Success Rates," Alta Mira Recovery Program, https://www.altamirarecovery.com/alcoholism/alcohol-rehab-success-rate/, accessed August 14, 2018.

CHAPTER 9

p. 201: *what with alcohol being the third leading preventable cause of death in the United States:* "Alcohol Facts and Statistics," NIH, National Institute on Alcohol Abuse and Alcoholism, https://www.niaaa.nih.gov/alcohol-health/overview-alcohol-consumption/alcohol-facts-and-statistics, last updated August 2018.

p. 201: *even one glass of wine per day can shorten a person's life by anything from six months to five years:* Meera Senthilingam, "Even One Drink a Day Could Be Shortening Your Life Expectancy," CNN, https://www.cnn.com/2018/04/13/health/too-much-alcohol-drinking-limits-shorter-life-expectancy/index.html, last updated April 13, 2018.

p. 205: *most racially diverse area of the United States and where the average household income is 34 percent lower than the national average:* "The Bronx," Demographics: 2013 Estimates, Wikipedia, https://en.wikipedia.org/wiki/The_Bronx#Demographics, accessed August 14, 2018.

p. 208: *risky drinking is rising fastest in the most marginalized communities:* Bridget F. Grant, S. Patricia Cou, and Tulshi D. Saha, "Prevalence of 12-Month Alcohol Use, High-Risk Drinking, and *DSM-IV* Alcohol Use Disorder in the United States, 2001–2002 to 2012–2013," *JAMA Psychiatry* 74, no. 9 (2017): 911–23, https://jamanetwork.com/journals/jamapsychiatry/fullarticle/2647079.